Scientific
Conversations

Scientific
Conversations

Interviews on Science from

The New York Times

CLAUDIA DREIFUS

With a Foreword by Natalie Angier

A W. H. FREEMAN BOOK

Times Books

Henry Holt and Company   New York

For Andrew

Times Books
Henry Holt and Company, LLC
Publishers since 1866
115 West 18th Street
New York, New York 10011

Henry Holt® is a registered trademark of
Henry Holt and Company, LLC.

Library of Congress Cataloging-in-Publication Data is available upon request.
ISBN: 0-7167-4661-1

Henry Holt books are available for special promotions and
premiums. For details contact: Director, Special Markets.

First Edition 2001

Designed by Cambraia Fonseca Fernandes

Printed in the United States of America
10 9 8 7 6 5 4 3 2 1

Contents

Acknowledgments

Enduring thanks go to Cornelia Dean, Editor of the *Science Times* section of *The New York Times*, who had faith that a wayward political journalist could make a path through the thickets of the scientific world. Much could be said about Cory Dean—her creativity, her openness, her loyalty to her people, but what is most striking about her is the model she casts of what a female news executive can be. Dean is at once professional and nurturant; she creates an atmosphere that is more communal than competitive. People give her their best.

Also at *Science Times,* gratitude is owed to my editors, Laura Chang, Barbara Strauch, and John Wilson—and to colleagues like Dennis Overbye and John Noble Wilford who helped me through my pledge period. Dennis in particular was the assigning editor on most of the pieces here, and his quirky ideas mixed with my irreverence, I think, made for a great combination.

At W. H. Freeman and Company, Erika Goldman had the vision and clout to commission this book and guide me into, I hope, making it into something more than just another journalistic collection. I am so grateful for her encouragement, her ideas—and also for the fun it's been to work together. Erika's assistant, Susan Goldstein, offered valuable help.

At *The New York Times,* Mitchel Levitas, Editorial Director of Book Development, was a marvelous in-house editor, adding skilled advice and great humor to this enterprise. His project coordinator, Tomi Murata, was gracious, kind, and wonderfully organized. Phyllis Collazo was kind enough to shepherd into print the many excellent photographs that enhance the book.

Thanks also to my literary agent, Deborah Schneider, to Maggie Pouncie who helped with the bibliography, and to Beatrice Dreifus, who is an unswerving source of love and support.

And finally, much gratitude to the Alfred P. Sloan Foundation for funding this book in part and for its objective of portraying the lives of the men and women who are engaged in science.

Foreword

Natalie Angier

Some time ago, while I was doing research for my first book, about genes that cause cancer, I brought one of the young researchers who would figure as a character in the book to a family dinner party.

"And what do you do for a living?" asked my grandmother, in her chattiest and most personable manner.

The researcher, Carmen Birchmeier, a woman from Germany in her early thirties, took a pull on the Marlboro cigarette that was never very far from her lips, let the smoke out slowly and said in her low, heavily accented voice, "I am a scientist."

My grandmother's eyes proptosed like a tree frog's. If Carmen had said she was a Tasmanian baronness my grandmother could not have been more befuddled. "A *scientist*?" she repeated. "You mean you study science in school?"

"I do scientific research in a laboratory," Carmen said, and briefly explained the particulars of her experiments.

"Well," my grandmother said. "I'm honored to have you here. I've never met a *scientist* before."

Throughout the evening, whenever Carmen opened her mouth to speak, no matter what the subject—my grandmother hushed up the table and leaned forward to listen. And when we were leaving, my grandmother pulled me aside to express her gratitude that I had brought such a personage into her midst. "She is the most intelligent woman I've ever met," my grandmother declared.

Okay, so maybe her reaction was a bit over the top—but not by much. Those of us who spend our lives talking to scientists tend to forget how rare and exotic a species they seem to the laity. To us, scientists are like the rest of us, a characterological potluck. But for many people, scientists remain beyond reach, comprehension or backslappability. They're not likely to meet scientists professionally or personally, and chances are they consider themselves scientifically illiterate, casualties of the high school chemistry or physics classes they practically flunked.

Moreover, our culture doesn't exactly flaunt its scientists. The most successful researchers in the world—Nobel laureates and similarly garlanded viziers of the profession—remain utterly unknown to the general public. And, as Leon Lederman, himself a Nobel prize–winning physicist, points out in this collection, we have yet to see anybody take the ultimate step in humanizing the research trade—by making a TV series out of it. Where is the drama, sitcom, or "reality"-based peep show chronicling the escapades of the laboratory rat pack? Where is *L.A. Science,* or *MIT 24/7,* or *The Neutrinos*? Sure, Hollywood occasionally serves up a scientist on the big screen—but that scientist usually dies or turns cackling mad before the audience has a chance to realize that the poor sod really is just like the rest of us. Small wonder that my family treated Carmen Birchmeier as though she'd just drifted in from the Oort Cloud.

Happily, with this volume of scientific conversations, Claudia Dreifus gives us not just one luminous dinner guest, but a whole blazing comet shower of them. In so doing, Dreifus manages the seemingly impossible: she brings her subjects down to earth, and humanizes and particularizes them, without in the least dimming their luster or clipping their tails. She doesn't fawn or gush, grandstand or fillet. Those of us who have written question and answer pieces know that the apparent simplicity of the format is deceptive, and that it is devilishly hard to get people to say interesting things well, precisely, and pithily. It is harder still to inspire somebody to join you on what amounts to an emotional and intellectual archaeology expedition, in which the source serves not only as willing dig site, but also chips in with the shoveling and sifting as needed.

And that means even, or perhaps especially, when it stings. Dreifus is not out to "get" anybody, but neither does she let her subjects get away with anything. For example, when questioning John Bancroft, the director of the Kinsey Institute for Research in Sex, Gender and Reproduction, about his prior use of "aversion therapy" in treating his homosexual patients, Dreifus asks him whether he regretted that part of his past. No, Bancroft insisted, because his motives for doing it at the time were "entirely honorable." Did he fear he damaged his patients? No, not at all. Waste their time, perhaps? Oh, yes, very likely. Does he consider homosexuality to be a sickness? No, not at all. Then why the aversion therapy? Well . . . the patients wanted it! They practically begged me to do it! And I can only thank heaven they don't ask me anymore! Get my drift?

Like any seasoned explorer, Dreifus has a kitpack of sturdy tools. One of her most reliably productive questions is the straightforward, "Give us a job description of . . ." an ethnobotanist, in the case of Michael Balick, or a theoretical astronomer, of Sir Martin Rees. But then there are the sweet surprises, like when she asks Rees, Do you ever get theorist's block? Or of Benjamin Carson, a pediatric neurosurgeon: How do you feel when you're digging around in people's brains? Or of Martin Wells, a zoologist who studies squid, cuttlefish, and octopus: You eat your subjects, don't you? Oh, yes, Wells replied. He and his wife have eaten calamari until they were "pig-sick" of it.

Some of the interviewees in this collection are as close to celebrity-hood as science types can be, mostly the result of their having written popular books—Stephen Jay Gould, Freeman Dyson, Sir Arthur C. Clarke, John McPhee, Sir Roger Penrose. Yet the book is less about name brands than it is about grand themes: the link between science and the soul; the purpose of music, beauty, and vaudeville; an ape's way of knowing, and an ape's sad way of dying; the know-nothingness of computers; the fate of the earth that we abuse with such flip abandon, and the revenge of an unfathomably vast universe that has yet to answer our piteous bleats of "Hey, anybody out there?"

I admit to a particular fondness for Claudia's emphasis on women: female scientists struggling to bring other women into the fold; women in medicine fighting against outrages like the widespread practice of removing young girls' clitorises and labia; Tanya Atwater, a geologist who responds to an earthquake by falling to the floor, the better to feel its every quiver; and Polly Matzinger, a former cocktail waitress, dog trainer, and Playboy bunny, who has shaken up the field of immunology with her radical notions that the immune system doesn't care a whit about "self" and "non-self," as the cardinal wisdom has it, but instead is designed to recognize danger whatever its origins. All told, female voices make up more than a quarter of this collection, which is an impressively high percentage given the reigning disparities in the scientific community.

This is a wonderful book, crackling with character, eloquence, quirk, and wit. You'll have fun reading it, and you'll learn a lot without knowing what hit you. And when you're done, be sure to pass it along to someone you love—like your grandmother.

Introduction

A CAREER SHIFT FOR CHRISTMAS

My journey into the scientific cosmos began at a Christmas party in the winter of 1997.

Till that moment, I'd been a political interviewer, globetrotting the world, witnessing revolutions and civil wars, questioning heads of state, experiencing the big prize that most journalists seek from their work—that famous front row seat to the events of one's time.

Indeed, in a thirty year career, I'd developed a strong reputation as a print interviewer. At the risk of sounding immodest, it was often said that my question and answer–style interviews shed new light on the psychological constructs of history makers and power brokers. Yet, despite these ego-strokes, in the winter of 1997, I felt restless in my work and unhappy with my political beat.

Perhaps I had sung the same song for too long? Or maybe I had just grown weary of the increasingly packaged characters who'd come to dominate public life in the late 1990s? Interestingly, I had fallen into this professional miasma at a time when most of us still associated the word "chad" with a country in west Africa, long before anyone had heard of Marvin Rich or Kenneth Starr, months before the kick-off of the most boring and insubstantial Presidential race seen in this lifetime. Though my own estrangement predates it, I think when history is written, the 2000 Presidential race will be looked on as a benchmark moment in the alienation of the American public from civic life.

But getting back now to that winter of 1997, I recall going to that Christmas party and feeling a kind of inchoate distress about my work life.

I remember feeling distracted as I made my way around the room and uttered holiday pleasantries to colleagues. I can recollect walking up to the buffet, being introduced to Cornelia Dean, the editor of the highly-respected Tuesday *Science Times* section of *The New York Times*. And I can recall Cory Dean saying to me, "Oh, Claudia Dreifus, I love your interviews! I'm thinking of adding an interview feature to *Science Times*. Might you be interested?"

From nowhere, I heard a disembodied and very flattered voice, *my* voice, exclaim, "What a lovely idea! I could do that!! I'd love to do that!!!"

And incredibly, I heard myself add, "Did you know, I have a bit of a science background?"

OHMYGODDDDDDDDDD!!! What had I done???

Though it was accurate to say that in the late 1970s, I had written on the politics of women's health, my work was more *politics* than health. And though it was true that I had an amateur's interest in nature and zoology, I was also the embarrassed possessor of the Mother-of-All-Math-Blocs. I'd failed geometry four times in high school, perhaps a world record. The reporters covering science at *The New York Times* were almost the scientific near equals of their sources. How could I—a woman who, until the advent of the calculator, couldn't balance her checkbook—hold my own with such giants as Malcolm Browne and John Noble Wilford?

On the other hand, a good reporter is a smart generalist, someone with the flexibility to cover almost anything under the sun—a phrase that happened to be literal at *Science Times*. Wasn't one of the key elements in journalism plain curiosity, anyway? "You can manage this if you pick mostly biologists and earth scientists," I said to myself, as I reported for my first assignment.

In the back of my mind rang out a mantra: "PHYSICIANS, YES!!! PHYSICISTS, NO!!!"

Of course, the very first scientist I was given to question was a physicist. In fact, the assigning editor, Dennis Overbye, an M.I.T. grad with a huge love of the cosmos, was rather partial to physicists. Thanks to him, over the next few months, I would learn that physicists came in endless categories—particle physicists, plasma physicists, nuclear physicists, theoretical physicists, more. Dennis would insist I get to know them all.

For openers, I was given the task of questioning Sir Martin Rees of Cambridge University, the Astronomer Royal of Great Britain, a physicist of the *astro*-variety.

Weak at the knees, terrified of failing, I prepared myself by taking a quick physics tutorial from *Science Times* senior correspondent, John Noble Wilford. I read, as best I could, Sir Martin's book, *Before the Beginning*, which I found surprisingly engrossing. Rees was a wonderful writer. I also went to a bookstore and purchased a copy of a comic book called, *Introducing: Stephen Hawking*, a kind of physics for dummies with fetching illustrations.

Martin Rees, I knew, was a close pal of Hawking's. Perhaps I could find something of use there?

When all was done, I constructed an interview that was more literature, psychology, and politics than science. My lead question, which was gleaned from a passage I found in a Passover service I'd attended the night before the session, was: "Nietzsche once wrote, 'If you gaze long into an abyss, the abyss will gaze back at you.' When Sir Martin Rees looks into the abyss, what does he see?"

In person and facing my tape recorder, Martin Rees was utterly startled by the off-centeredness of my query. He'd been interviewed dozens of times over his long career, but who on earth was this weird woman *The New York Times* had sent?

Rees hesitantly smiled. It seemed for a moment that he understood what I was trying to do—which was to get playful about science. And bless him, this dear genius was willing to climb into the sandbox with me.

For the next two hours, we tossed about notions on D. H. Lawrence, extra-terrestrial life, Stephen Hawking, and yes, even astrophysics. I even managed to use my lack of expertise with this little question, "When it comes to astrophysics, most of us are perplexed because the cosmos seems too complex to understand. Why should the ordinary Joe or Jane know their astrophysics?"

As a *coup de grace*, I ended the interview with the dumbest, fun-est question I'd ever asked a source: I inquired of the Astronomer Royal his astrological sign. You'll have to jump to page 7 to see his response.

A few days later the piece appeared, and amazingly, it was a hit. Friends telephoned with their congratulations. Fan mail arrived. A former *Newsweek* bureau chief, someone I'd never met, wrote that he knew "how hard" it was to do what I do.

Keer-ching!!! I had myself a beautiful new career.

It turned out that my outsider status to the culture of science was a plus; it gave me the chance to be a kind of medium for the reader with hard-to-grasp concepts. I didn't come into interviews with a lot of baggage. And in science, as in politics, there exists the counterpart of ideology.

As a newcomer to the field, sources didn't have any particular notions about who I was and what I thought, and so they distorted themselves less to please me than they might have with a science insider.

Moreover, the procedural reality that every time I faced a new topic, I needed to teach it to myself meant that I was an excellent translator for difficult ideas. In order to "get it" myself, I had to break things down to their simplest level.

And then there was another bonus. Scientists, unlike politicians and film stars—my previous quarry—had not, for the most part, been over-interviewed. More often than not, they came to an interview without a posse of professional handlers, but with great unheard stories to tell. In a era when Jennifer Lopez's outfits are the stuff of headlines, the media had mostly ignored this crowd. My science sources were not spoiled.

One of the cardinal rules of interviewing is to try to pick subjects who actually want to talk. With a science beat, I now had a whole field full of virgin subject matter to explore. All this freed me to be far more creative, I believe, than I've had the chance to be before.

Interviewing, to me, is an art form—but it is one where both sides of the process must be willing to perform. I am more of a *developmental* than a confrontational interviewer; I prefer to like the people I write about and through a process of exploration and empathy, extract their stories from them. On the science beat, I'd hit interviewer's heaven.

COVERING A *REAL* REVOLUTION

In the time since that Christmas party, I have produced some fifty science interviews, thirty-eight of which are collected here—with additional postscripts on what these good people have been up to since our encounters.

In my years as a political journalist, sources rarely became friends. And that's probably a good thing. But on science, you'll see, many of the inter-viewees stayed in touch; several became pals. Finding new friends has been one of the pleasures of this work. There have been others, too.

By leaving the world of politics, I was astonished to discover that I was getting a chance to witness a *real* revolution. Over the years. I'd reported on the upheavals of my time. I was in Northern Ireland in 1969 when a civil rights campaign exploded into the vicious civil conflict still known to the world as "The Troubles." I went to Nicaragua in the 1980s, and to Chile in the winter of 1990, when an election pushed a dictator out of the presidential palace and opened the door to the redemocraticization of that wounded nation. But on the fourth floor of *The New York Times* building, the place where the science section of the paper is produced, I've witnessed an extraordinary amount of real social and, ultimately, politi-cal change.

Think of this: In the time I've been working in science, Dolly the Sheep was cloned, the Genome Project's completion was announced, signs of water on Mars were photographed, new planets were discovered, the Internet became ubiquitous, and many of the mysteries of Alzheimer's were untangled.

And within science itself, there has been an internal revolution to observe: the changing face of who gets to do the research. In 1970, when I was just out of university, the number of women in science was at 13.6 percent; twenty years later, it was 33.5 percent. And growing.

To get a sense of how this revolution happened, take a look at Eleanor Baum's interview on page 229. Dr. Baum, one of the few female engineers during *The Feminine Mystique* era of the 1950s, has used her position as dean of the prestigious Cooper Union School of Engineering to transform her profession. Baum's experience shows how a committed leader and some fearless affirmative action can drastically alter societal notions of what men and women can do.

Idealism was what drew me to politics long ago, and I'm happy to report that I've encountered some amazing idealism in the world of science. Here's Rita Colwell, microbiologist and the first woman ever to head the National Science Foundation, explaining to me how she used her inventiveness and common sense to save thousands of Bangladeshis from cholera:

I thought if one could remove the zoo plankton—the copepod on the zoo plankton serve as host to the cholera bacteria in the drinking water—we could go a long way towards curbing the disease. The problem was that sophisticated water filters were too expensive for Bangladesh, one of the least wealthy countries on earth. So I thought, "What could be used as a filter that exists in daily life?" The answer was cloth, sari cloth, which even the poorest of the poor have.

One of the things I came to abhor on my old political beat was how packaged most politicians had become. In the 1960s, when I first began writing, public life was full of vivid characters. In the U.S. Senate alone, there were giants like J. William Fulbright, Barry Goldwater, Robert F. Kennedy and that constitutional curmudgeon, Sam Ervin. In this era of sound bite wariness, most officeholders are so cautious that there's rarely a point to a question-and-answer-style interview. The openness to make a Q. and A. successful is, mostly, absent.

But scientists-as-subjects were certainly not overinterviewed and they were certainly not prepackaged. The entire field—from astronomy to zoology—was chock-full of quirky individuals who had no problem with being themselves. Artificial intelligence guru Marvin Minsky (page 97), met me at the door of his Boston home wearing a shirt festooned with masking tape. Primatologist Emily Sue Savage-Rumbaugh (page 83) lived with her family of bonobo apes at their facilities within her Georgia State University Research Station. The great mathematician Sir Roger Penrose (page 143) had Lego toys strewn about his office at Oxford University. This gentle genius *liked* playing with them.

The best part of my new profession came in the political year 1999–2000, a time when most political journalists were overcome with self-loathing as they were forced to write about George W. Bush's drinking habits and Bill Clinton's sex life. Typing away at the *Science Times* for that entire presidential campaign, I didn't have to utter the Monica-word once, unless, of course, I wanted to.

And in the summer of 2000, I finally did. I was in Oakland, California interviewing forensic mathematician Charles Brenner. Dr. Brenner's claim to fame was that he did probability work with DNA testing, and so it seemed appropriate to ask him, "What did you think when you first heard of Monica Lewinsky's blue dress?"

"I tried not to pay attention," Dr. Brenner replied. "I was impressed by Bill Clinton's toughness in the situation, but it did strike me that if you could get his typing, his goose was cooked."

It's this kind of plain speaking that makes me think that an interviewer's paradise is on floor four of *The New York Times* building.

SOME RIFFS ON TECHNIQUE

In my 1997 book, *Interview,* I wrote extensively about the zen of interviewing. Except for the subject matter, interviewing techniques are similar in politics, culture, and science. Some of my methods may sound obvious. But in journalism, as in cooking, the simple can be sublime. One of the great journalists of the twentieth century was a *Timesman* named Homer Bigart who, in the field, asked questions with the ingenuity of a child.

My basic rules for any type of print interview: pick an interesting interviewee who wants to talk, learn your subject matter as well as you can, prepare a line of questioning in advance, but don't necessarily stick to it. The vital thing is to stand back and let the interviewee do the talking.

The most important decision interviewers make is in picking of a subject. Because in a Q. and A., the words that are elicited are what ultimately makes up the body of the article, finding an articulate source is key.

This may seem terribly obvious. But in Q. and A.'s, a journalist does not have the saving option of filling out a disastrous interview with interesting reporting. If the discussion on tape is sparse, the interviewer will go home empty-handed—a situation highly displeasing to editors.

This fact-of-life can lead to some heavy-handed triage in the choosing of subjects. But I'd say: In most cases, if a source has a reputation for being a poor storyteller, if they are known to be reticent, or to talk prepackaged sound bites, it's best to pass on them.

How can an interviewer sense who will be interesting? If the potential subject is an author, I'll read their books. Are they chatty, accessible—do they sound as if they'd be open to an interesting line of inquiry? Do they seem . . . stuffy?

Sometimes, I'll watch *NOVA* on PBS or listen to NPR's *Science Friday* to see how a particular scientist handles questions. I'll also look into *The New York Times* archives to find previous interviews with the person and then go on-line to see what Internet has on them. The advent of the Internet is the best thing to happen to journalism since movable type. I often pick up bits of factoids that I would never have found by the old methods of a decade ago.

I rarely choose subjects that university public relations officers pitch me. I tend to veto potential interviewees who churn out curriculum vitae that exceed twelve pages—it's a sign of insecurity and egotism. For the same reasons, I'd like to, but often cannot, pass on academics who have more than twelve words in their official titles.

Of course, sometimes one just can't tell much in advance. Perhaps the worst interview I've conducted ever was for *Science Times*. It is not reproduced here because it never appeared in print. The talk was with an elderly Scandinavian naturalist who'd written a charming memoir about his life and work. Alas, when I arrived for the interview session, he was plagued by an attack of Nordic reticence so strong that he wouldn't even tell me his opinion of the theory of evolution. I also won't mention here the world famous geneticist who arrived in my hotel room for a tape session stark raving drunk. That interview didn't run either.

The second most important step an interviewer can take is in doing his or her homework. I try to prepare the way a Ph.D. candidate might for their orals. I read everything about the subject that I can get my hands on. I look at competing works for ideas about alternate theories or practices or both.

Though it's gauche, I'll phone up a would-be source and see how they fare in a dry run over the telephone. Moreover, I may ask the subject to give me the names of close friends or colleagues so that I can do a preinterview with them. Often I ask them, "Tell me something about Dr. So and So that no one knows about her." Then, I'll fashion a question from this tidbit of intelligence.

I'm always scavenging about for interesting questions to toss at my people. I search for them with the same zeal I use in routing through flea markets. If, in my reading on philosophy or poetry or cooking, I find an interesting quote that could relate to some area of science, I record it and put it away for future use. Don't be surprised if references to such humanities folk as Hannah Arendt and Muriel Rukeyser turn up in future *Science Times* interviews. To me, a good interview should be a little bit surprising for the reader, as well as the interviewer. It helps to mix the disciplines.

The first question I ask in the actual interview session is critical. It sets the tone for everything that will happen subsequently. It shows that I'm serious, that I've done my preparation, that I've thought a lot about the subject and his or her work. I often spend a huge hunk of my preparation time on fashioning a lead question that I hope will create some good ignition.

I'll usually dress up for the interview. It's a way of honoring the process. I always try to bring two tape recorders with me. One machine will invariably fail—no matter how expensive it was. Of course, with Murphy's Law being in constant operation in this cosmos, I pretest my equipment and carry spare tapes and batteries—just in case.

About what happens in the interview session itself: some of it is magic. Don't ask me to quantify it. What can be said is that successful interviews are about being a good listener—about the spark of conversation and ideas, about the chemistry of personalities. I'll have my bag of questions, but I'm always willing to stray from them.

Very often, I'll ask to do the interview in a setting that the source is comfortable with, but one where they will not be posturing—for example, their office or their laboratory. I best like to interview people in their homes. They'll be relaxed there and I'll also find clues around—artwork on the wall, books in the library, photographs on the mantle—that can lead to a revelatory insights. Sometimes—and this usually works well—an interview will take place in the field. I interviewed ornithologist Luis F. Baptista (page 57) in San Francisco's Golden Gate Park while he described the soap opera life of his local friends, the sparrows, to me.

When I return to my desk, I'll transcribe the tapes myself. Though this is tedious and no doubt an invitation to carpal tunnel troubles, it does give me a sense of the subject's language rhythms, intonations, and what they actually mean by a specific phrase. It's important to leave real language in; one wants to hear distinct voices. Later, I'll rout through the finished transcript, eliminate dull or repetitive sections, find the spine of the piece, and pull it all together.

When it comes to the writing stage of the interview, I like to think of myself as something like a playwright. In effect, I am creating a two-person play—where the journalist is the minor character. I try to use my questions to move the interview along, not to show the readers how clever I am.

The tendency for interviewers to use their stories to show themselves off is one of the lamentable results of undisciplined interviewing. Not every journalist is an interviewer. It's a skill that requires training, tact, and most certainly, restraint. The reporter who can't stand back shouldn't do it.

At the same time—and this is not contradictory—interviewing is a part of journalism that requires that the reporter use more of his or her personality than other types of work. When I walk into an interview, I am bringing everything I am—my personality, my education, my ideas, my temperament, and my life experiences. I am looking for a kind of intimate connection with my subject—something like transference, and I'm looking to establish it quickly.

Frankly, that interviewing is such an individualistic form of journalism is gratifying. Decades ago, as a drama student at New York University, I studied playwriting and acting; I get to use these skills every day in my life as an interviewer. Indeed, when I write the short introductions to each Q.&A., I put them together as if they were stage directions at the beginning of a play.

But back to science interviewing. At the end of the day, what makes science interviewing such a blast is how marvelous the people are and how many of the important changes for our lives and societies in the twenty-first century will, very likely, come from them—revolutionaries, indeed.

Indeed, in this year of 2001, a good political journalist must, absolutely, know her science. As I write these words, just about every major policy issue that the administration of President George W. Bush is confronting has a science component to it—global warming, fetal tissue research, stem cell therapies, the proposed anti-missle defense system, the possible resumption of nuclear arms testing, oil drilling in the Alaskan arctic. By moving onto a new beat, I've recovered my old one and it is a privilege to have gained the knowledge that permits me to be a part of the debate.

Often, my new speciality fills me with awe. Think about it: If we are lucky, Alzheimer's and other dementias will be history, as will many birth defects and cancers. With more luck, we may solve food problems on earth and begin seriously colonizing the heavens. Some of the people working at these miracles are presented here. I'm hoping you have as much pleasure meeting them as I've had.

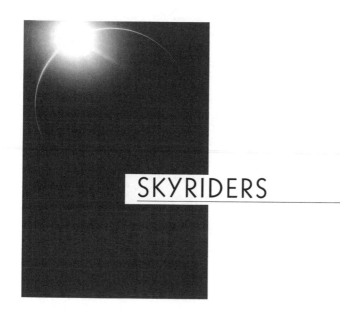

SKYRIDERS

SIR MARTIN REES

Tracing the Evolution of the Cosmos from Its Simplest Elements

Sir Martin Rees, 55, one of the world's leading theorists on cosmic evolution, the Astronomer Royal of Great Britain and author of *Before the Beginning: Our Universe and Others* (Addison-Wesley, 1997), was in the United States this month to receive the Franklin Institute's $250,000 Bower Award for "achievement in science." Sir Martin spoke with *Science Times* over lunch in the cafeteria at the Institute for Advanced Study in Princeton, New Jersey.

— *Nietzsche once wrote, "If you gaze long into an abyss, the abyss will gaze back into you." When Sir Martin Rees, Astronomer Royal, looks into the abyss, what does he see?*

First off, I should mention that Astronomer Royal is a purely honorary post, rather like Poet Laureate. The Astronomer Royal's duties are so exiguous that they could be performed posthumously. Therefore, I have a day job as a research professor at Cambridge where I interpret astronomy data, rather than observe. Other people do the gazing.

What I do is to try to understand how our universe has evolved from simple beginnings to the complex cosmos we see around us, of which we are a remarkable part ourselves.

— *When it comes to astrophysics, many of us are perplexed because the cosmos seems too complex to understand. Why should the ordinary Joe or Jane know their astrophysics?*

Because there's a fascination with our origins and astrophysics is the key to it. If we are to understand an everyday question like "Where did the atoms we are made of come from?" we must understand the stars. Did the creator magically turn 92 different knobs to make the different elements? Or is there a reason why the earth contains a lot of carbon, oxygen, and iron but not much gold and uranium?

3

The explanation is that all the atoms were once inside a star. When our Milky Way Galaxy was first formed about 10 billion years ago, it contained the simplest atoms: hydrogen and helium.

Then, the first stars were formed and the nuclear fuel that kept those stars shining converted hydrogen into helium through nuclear fusion and then converted helium into other atoms: carbon, oxygen, and the rest of the periodic table.

Later, the stars ran out of fuel, they exploded, threw back all that debris into interstellar space and it all eventually condensed into new stars. One of which was our sun.

— *So when the poets sing, "You are the sun and the moon and the stars," they are being literal?*

We are the dust of long dead stars. Or, if you want to be less romantic, we are nuclear waste.

Sometimes people ask me, "Are we presumptuous to think we can understand anything as big as a star, or a galaxy, or the Big Bang?"

The response I give is that what makes things hard to understand is not how big they are but how complicated. Inside a star, everything is broken down to its simplest constituents. Ditto, in the Big Bang.

On the other hand, it is much, much more difficult to understand the simplest living organism. The most wonderful thing we know about in the universe is life, and that's the most complicated emergent phenomena we know of.

I'm always amazed when we study these simple beginnings, one has not just understood how the chemical elements have been made but how they've forged themselves into something complicated enough to develop into life.

Another motive for studying the cosmos is that it is a cheap way to understand and test the basic laws of nature under extreme conditions. We can't simulate strong gravity or the physics of the Big Bang in our laboratories. So astronomers can complement the knowledge gained by scientists on earth about the fundamental laws.

— *Getting back to Nietzsche, how do you think "gazing long into an abyss" changes a person?*

It gives one a slightly different perspective on time scales. From astronomy, one learns the immense time spans involved in cosmic evolution—billions of years. And for me, a much more important realization is that we are still at the beginning of cosmic evolution, not the culmination. Even our sun is less than halfway through its life. That makes me feel we should regard ourselves as part of the natural order, rather than the culmination of it.

What was it that D. H. Lawrence wrote? "I am part of the sun as my eye is part of me. That I am part of the earth my feet know perfectly and my blood is part of the sea." And that understanding leads to the question, "Is life as we know it unique to the earth?"

The origin of life is a harder problem than most astronomical problems and therefore you don't really have enough evidence to say whether the evolution of life is likely or unlikely. One of the developments in astronomy in the last two or three years is that we've realized that there are certainly planets around many other stars and almost certainly many of these planets have the environment of the young earth.

What we don't know is whether given the right conditions, the emergence of life is automatic or a rare accident. Still less do we know whether simple life automatically evolves toward something we recognize as intelligent. So we don't know whether life is unique to the earth.

— *Do you favor projects searching for life elsewhere?*

Absolutely. Either for simple life in our solar system or intelligent life elsewhere, or just for signals of intelligent life, because even if these signals didn't make much sense to us, if you found a signal that was clearly artificial in origin, that would have an impact because it would show that concepts of logic and physics weren't restricted to the hardware inside of human skulls.

— *If you had to guess how the search would come out, how would you bet?*

I think it's unlikely that there's intelligent life elsewhere, but I'm enthusiastic about searching for it anyway. Actually, I would be happier if there weren't any life out there. If life is unique to the earth, it gives the earth—small though it is—a real cosmic significance.

Moreover, if one also recognizes that our earth and our solar system are nowhere near the culmination of their evolution, then we could imagine that species descended from us could propagate out through the galaxy.

There's as much time ahead of us as there has been in the course of evolution until now. So on that time scale, life could spread from the earth. I think this uniqueness would give people an extra motive for ecological concerns in the sense that if life were to be snuffed out on earth, we may be destroying genuine cosmic potentialities.

— *You're an astronomer-theorist. Can you give us your job description?*

Well, there are some theorists, and Steve (Stephen W. Hawking) is one, who work in a very mathematical way, constructing systems that may be rather remote from any observations. I'm someone who's very closely in touch with data and phenomena. Most of my work tries to tie together work done by astronomers with telescopes and spacecraft, to make sense of it, to see if it falls into some kind of a pattern.

— *Do you ever get theory-block?*

Yes. And the way I overcome it is to work on several things at once. I follow several areas of a subject in parallel. I'm on three or four at the moment. I have a rather short attention span and if one thing goes badly, then I shift to something else. Also, I often explore at the same time two contradictory interpretations of a phenomena.

— *There's a famous quote about your profession, "A theorist may be wrong, but never in doubt." Have you ever been wrong?*

Oh, often. If you're never wrong, you're not being bold enough. Obviously, when some new phenomenon is discovered, you have limited data. It would be astonishing if you guessed right the first time. I think you have to accept that all our knowledge is provisional.

When I started in the 1960s, I thought it was the most exciting time ever. The first quasars, the radiation left over from the Big Bang, the first pulsars and neutron stars were discovered. And that was very good for young people because it meant that the experience of our elders was at a heavy discount, and we could all start explaining things from scratch. Now what's been exhilarating is that the rate of discoveries has not diminished.

If I were to think of the last two years, there have been just as many remarkable discoveries as in any earlier period: evidence of planets around other stars, new evidence on the very distant galaxies, more ideas on the Big Bang. There's just as much happening as ever.

— *In recent weeks, there have been exciting claims that the universe's expansion is accelerating. What does this mean for astronomical theory?*

We are trying to decide if our universe will continue expanding forever, or if the firmament will eventually crash together in a big crunch. The evidence strongly favors perpetual expansion.

Still more exciting is the claim that the expansion is actually speeding up. This would imply that there was a repulsive force even in empty space, as was envisaged by Einstein when he postulated a so-called cosmological constant. The jury is still out on the cosmological constant. Whether this constant is zero or nonzero is of fundamental interest to theorists, and the answer will help forge another link between the microworld and the cosmos.

Many cosmologists hope it's zero because they're prejudiced in favor of the "simplest" universe. But I'm relaxed on the issue. I believe an infinity of different universes actually exists, and we could find ourselves in any one in which the governing laws allowed life to exist.

— *How does someone who theorizes about the cosmos feel about a phenomenon like astrology?*

I don't know any astronomer who takes it seriously. The question is, should one campaign against it? If I were an East Indian, where the leaders subscribe to astrology, I would campaign against it.

Maybe if I was an American in Reagan's time, with Nancy, I might do so also. But in Britain, we have, I think, a more sophisticated attitude than in U.S. popular culture.

Neither creationism nor astrology are serious issues. So I don't really feel that it is appropriate to take a very strong line. Indeed, if one did so, one would be perceived as unduly solemn.

— *So what's your sign?*

Cancer.

— *And what does it mean to be a Cancer?*

I wish I knew.
April 28, 1998

Postscript

Sir Martin Rees, my first *Science Times* interview, is one source I still hear from via e-mail.

Here is his February, 2001 missive:

"Probably the only update worth mentioning is that I have two further books: *Just Six Numbers* (Perseus Books, 2000) and *Our Cosmic Habitat,* (Princeton, Fall, 2001).

"Yes, I am still Court Astrologer! Research on same topics: plus on 'cosmic dark age' before first stars were formed and on planetary systems."

SIR ARTHUR C. CLARKE

An Author's Space Odyssey
and His Stay at the Chelsea

Edward Keating/*The New York Times*

On a recent autumn morning in a suite at his beloved Chelsea Hotel, legendary New York City home of hipsters and artists, **Sir Arthur C. Clarke,** 81, the author of more than a hundred books of essays and science fiction, co-writer with Stanley Kubrick of the film, *2001: A Space Odyssey,* and the dreamer who in 1945 came up with the idea of the "comsat," or communications satellite, held court in a wheelchair tended by his two Sri Lankan valets, Hector and Lenin.

The author, who lives in Colombo, Sri Lanka, had stopped off to see old friends and admirers on his way home from a round of medical tests at Johns Hopkins Medical Center in Baltimore.

— *Is it true that you and Stanley Kubrick wrote the screenplay for* 2001: A Space Odyssey, *right here at the Chelsea Hotel?*

Absolutely. This place is my spiritual home. Everyone is surprised that I come to this hardly five-star hotel.

In fact, the Sri Lankan ambassador was horrified. Before coming here, I had been at Johns Hopkins for more than a week.

They took a nerve biopsy. I have some obscure neuropathy. All of a sudden, my legs get numb. Mainly, I have post-polio syndrome, which is why I can't walk anymore. I was told I had Lou Gehrig's disease, originally.

As far as I know, Stephen Hawking is the only man who's survived that for very long, so this (laughs) is a considerable improvement.

— *We heard that the astronaut Buzz Aldrin made a visit to you at Hopkins.*

Yes, he dropped in to see me in the hospital, and he kept making the point: We've got to get NASA out of the space business. He believes it should be private enterprise. There are a lot of people now who are trying to develop relatively cheap nonbureaucratic access to space. How successful they'll be, I don't know. I think the rocket will end up doing for space travel what the balloon did for air travel: It got us there, but soon was superseded by something better.

— *Are you as critical of NASA as Buzz Aldrin is?*

I won't criticize NASA because it's the slave of Congress.

Actually, NASA has now set up an Institute for Advanced Concepts. It is looking at all sorts of crazy ideas, including my favorite one, the space elevator. It's such a delightfully simple idea: build an elevator from the equator to a geostationary satellite. You move payloads up and down by electricity.

When I wrote my book *The Fountains of Paradise in* 1979, about the building of the space elevator from a country which is 90 percent Sri Lanka, the only material that could be strong enough to build a space elevator was diamond. And that, unfortunately, was not available in the megaton quantities needed for such a project.

Interestingly, when I later recorded a 12-inch L.P. of *The Fountains of Paradise,* the cover notes and illustration were by Buckminster Fuller.

And now, we have the material to make it: C-60 nanotubes, which are hundreds of times stronger than steel.

The Rice University scientists who were co-discoverers of C-60 have named it Buckminsterfullerene. If C-60 can be mass produced, it will revolutionize space travel, as well as everyday life. You could lift your car with one hand.

The point of the space elevator is that it makes space travel cost just pennies. The cost in electricity of lifting you to space is about $200. The cost for a round trip is about $40 because you get most of your energy back on the downward trip. I've said many times that the main cost of space travel in the future is going to be for catering and in flight movies, not for fuel.

And of course the considerable interest on the trillion or so that it costs to build the thing.

— *One reason you advocate space travel is fear of asteroids?*

I'm always quoting the science fiction writer Larry Nivens that "The dinosaurs became extinct because they didn't have a space program." Yes, I think a serious asteroid impact is inevitable. In any case, even without them, we have all our eggs in one basket here. By the way, I'm an absentee landlord of a hundred square miles of some rather rugged territory near the orbit of Mars. I have an asteroid named after me. Isaac Asimov's got one too. It's smaller and more eccentric.

___ *In the news is word that the world's population has hit six billion. What do you feel when you see a headline like that?*

Well, I feel rather depressed, but then there are so many times when I'm an optimist. I think we have a 51 percent chance of survival. I would say the next decade is perhaps one of the most crucial in human history, though many people have felt that in the past. But it's real now. There are so many things coming to a head simultaneously. The population. The environment. The energy crunch. And, of course, the dangers of nuclear warfare. I am often asked to predict things and I'm described as a prophet, but I deny that. I'm just an extrapolator. I can envision a whole spectrum of futures, very few of which are desirable.

But I do feel that we science-fiction writers do serve as an early warning system, by showing what may happen, what could happen, and simply what should happen. I've often said one reason why I'm an optimist is that then you have a chance of creating a self-fulfilling prophecy—and ditto if you're a pessimist—it's more cheerful.

___ *Can you think of something that you predicted might have happened that doesn't look as if it will?*

Well, there won't be a Hilton in orbit by 2001.

___ *Was it your idea or Kubrick's to make a computer, HAL, the villain of film?*

Villain? HAL's a nice guy. No, I cannot now say who did what. The only thing I'm completely sure of is that the idea of HAL lip reading, which I thought was rather unlikely, was Stanley's. Now, of course, they've succeeded in getting a fair degree of accuracy with lip reading in computers. The way we wrote it—we just brainstormed. We sat in this apartment and brainstormed.

Stanley said, "I want to do a good science-fiction film." So we went through all my short stories to see what would make a good film. We had about six. One by one, we threw away stories. Eventually, we just had two of them. One was *The Sentinel,* and the other was *Encounter in the Dawn,* in which a space ship lands before man existed and the travelers meet man-apes. We were originally going to call the film, *How the Solar System Was Won.*

— *One of the legends about you is that you came up with the idea for Comsat in an article you wrote in 1945 and that you never patented the idea.*

Oh, so you want to ask me about how I lost a billion dollars in my spare time? Well, you see when I wrote my "comsat" paper, it was 1945. The war was still on. No one could even imagine what peace would be like. And I didn't think that satellites could be launched until the end of the century. So I didn't give the matter of a patent any thought at all. I just wrote this article and sent it off and got $15 for it. Which was real money in 1945. I don't regret it because I think a patent would have expired anyway before the first comsats were launched. Until the technology reaches some level, you can't patent anything. Besides, who was it that said, "The patent is merely a license to be sued?"

— *Surely you must sometimes fantasize about what your life would have been if you had?*

O.K., what I should have done is to try to copyright the word "comsat." If I'd done that . . .

— *You've written somewhere near 100 books of science fact and fiction. How fast do you write and think?*

Not as fast as I used to. I have difficulty remembering names. But I feel as long as I can spell "Alzheimer," I'm in good shape. I don't know if I've ever written more than 3000 words a day. Isaac Asimov could do that in a minute and a half. And mostly on a manual typewriter, too.

— *There are plaques mounted in front of the Chelsea Hotel dedicated to deceased artists who once lived here: Thomas Wolfe, Brendan Behan, Virgil Thomson. Will there be one for you one day?*

Oh, I hope so, but not for a long time. However, I don't mind if they put one up right now.
October 16, 1999

Postscript

Sir Arthur C. Clarke was not well on the September morning that I visited with him in his shabby-chic suite at New York's Hotel Chelsea.

He suffers from something that must feel as awful as it sounds: "Demyelinating Polyneuropathy."

Yet, the man had arranged to have the absolute best time a sick person away from home could possibly invent. His rooms at the Chelsea were a kind of court, where the prominent and interesting came to pay tribute. My interview was sandwiched between homages from Woody Allen and Rupert Murdoch. Who knew that Rupert Murdoch made sick visits?

And so, it wasn't surprising in March of 2001 when I e-mailed Sir Arthur in Sri Lanka for an update on his recent activities, to get a note back from Sunil Amarasekara, his secretary for e-mail, that said: "Thank you for your mail to Sir Arthur—he is in very poor health at the moment but has asked me to send you the attached Egogram: He is looking forward to your book."

"The Egogram" is a Clark-esque form letter he wires out in the same way ordinary mortals send a Xeroxed Christmas note with their holiday greeting cards.

Here are excerpts: "2001 opened with an explosion of film-related publicity— at one time I was receiving fifty e-mails a day from people wanting interviews etc. Thank goodness it is now over—and how sorry I am that Stanley Kubrick is not here to enjoy the praise he deserved. Though I am feeling fine, I am completely wheel-chaired, which does not stop me from playing an illegal form of table tennis, leaning on the table, so that my serves completely confuse any hapless opponent. Well, the Minister of Science (of Sri Lanka) has announced that the Cabinet is allocating $20 million for the development of the Arthur C. Clarke Technology Township. My greatest hope is that it will contribute to the economic growth of Sri Lanka and thus to its peace and stability."

F. STORY MUSGRAVE

Watching from Sidelines as NASA Regains Spotlight

Joe Skipper

It was a sweltering recent autumn afternoon and **F. Story Musgrave**, 63, physician, poet, skydiver, and now former astronaut, was sitting in a Florida hotel room and fiddling with a recalcitrant air-conditioner. "I'll get this going in a flash," declared Dr. Musgrave, the former space walker. "I'm real good at fixing things."

Indeed, he is. In 1993, when the NASA needed someone to fix the near-sighted Hubble Space Telescope, Dr. Musgrave was the successful mission commander. Three years later, he was back in space, deploying and retrieving satellites for the study of the origin and composition of stars. In all, he has flown on the shuttle six times, more than anyone else. Last year, under conditions not entirely of his choosing, Dr. Musgrave retired from the space program. He explained why he left his "calling" in September 1997, after the air-conditioner kicked on.

___ *Why did you leave the space program?*

They told me in the most unambiguous terms that they weren't going to fly me anymore. I could have fought it. Lots of people urged me to. But I just figured that space flight is a huge privilege, and I've had an incredible career.

___ *Do you think you were eased out because of your age? You were 62.*

I don't know. I do think NASA is uncomfortable with people who make a career of being an astronaut. For most astronauts, NASA is a stepping stone to other things. I think they get uncomfortable with someone who believes, "Space is it, and I put it ahead of institutions." I had done that. I always said, "Space is my calling." My own guess is that this was as important as my age.

___ *You were the oldest of the astronauts. Were you a good subject for experiments on aging and space?*

15

Oh, yeah. Agewise, you can't miss with me. If you want to make some conclusions about age and space, there are decades worth of scientific data on me in NASA's own format.

Also, I can fly as a full-fledged crew member who does everything, not just a body they are using for study.

The thing I learned about aging and space is that the more I did, the easier it got. Whatever functions you lose because of aging, you gain in terms of experience. I remember that after landing from the Hubble repair mission, I did five hours of medical exams, got on an airplane, flew home and went to work the next day at 8 A.M. I made the circadian shift in one night. My last shuttle flight was even easier. I walked off and was steady as a rock.

— *NASA will be flying John Glenn, 77, into space soon. Are you feeling a twinge of envy about his being given the mission?*

I feel we should fly him. He was the first American into orbit. It's probably the right thing to do in terms of historic closure. And I feel that the reasons he is flying, and I'm not, are probably not related. The real reason he's going up is because he's John Glenn, and he's a Senator, and he convinced them to fly him. After they made that decision, science—looking at issues of aging in space—became the rationale for justifying it.

Actually, I think it will be difficult to make many scientific conclusions from Senator Glenn's flight. What they are doing is taking a John Glenn, who has led the life of a Senator for decades, and all of a sudden, switching his life to that of an astronaut. Now, what effect does that have? Are you really studying space flight? Due to the multiple factors involved, such as the total change in the man's life style, it will be very difficult to attribute any observed changes to age alone.

The other thing that bothers me about the Glenn flight is that he's chosen to do his space part-time, while still a Senator, while still living in Washington. We all know that you get out of a space flight what you put into it. I do wonder if he'll have much of a space experience, other than just the survival training.

— *What's a "space experience"?*

Two principal things: what you see out the window and also your cognitive sense of your mind and body in terms of the free-fall condition.

When I go up, I give myself over to the space experience, surrender to it, let it touch me. I'm always saying to myself: "Story, look around. Don't just go out there and fix the Hubble, look around."

___ *A more earthbound question: Are you critical of the way the space program has been administered?*

Listen, NASA's a magnificent organization at the technical level. I have such an incredible pride and love for the people who work space—not so much for the leadership.

We need to assume that "the Big We"—the President, the Congress, NASA, the aerospace industry, the public—right now cannot manage any space program. Think about it: We have spent $20 billion over 14 years for a space station, and we do not yet have a screw in orbit!

Another serious problem is that launch costs are too high. The most important priority should be to have low-cost reliable access to space, and we seem to have no interest in that. I worry that we have not developed a new launch vehicle in over 40 years. We're still flying *Atlases* and *Titans* and *Deltas* from the 50s.

___ *Do you think the space station will actually happen?*

I believe so. We are into it for $20 billion and 14 years. It would be extraordinarily difficult to turn away from that kind of investment. The current posture we are in, however, is absolutely desperate.

The posture that bothers me with the shuttle costing more than $600 million a flight and with our flying this very old, very fragile technology, is an accident could happen and the whole deck of cards will come down. You can't do the space station without the shuttle, and the shuttle is old and fragile.

Another problem with the current posture is that the space station will dictate resources and make them dedicated to earth orbit activities for the next 10 or 15 years. Also, I worry that it's all going to be so expensive, so bureaucracy-bound, that we won't be doing cutting-edge projects anymore. With this current posture, we may not pursue that science which bridges cosmology and philosophy—the science that gets at who am I and what is the meaning of all this. We must do science that touches to the core—which means science that asks, "Who am I?" Anything less does not excite the public.

— *A more earthbound question: You've recently moved your homebase from Texas to Orlando. Why?*

I had a vision for decades that when something brought an end to my calling, I would get into a car and drive to Hobby Airport and go to Orlando. That's as far as the vision went. I never saw anything else. Then, when I decided to leave NASA, Disney asked me to be involved in their work. So the vision kind of came true.

Right now, I'm a consultant to Disney's various projects. I work on creative ideas for new resorts, new parks, new pavilions. I'm consulting on movies. One of my own personal goals here is to find ways to communicate the experience of space to people without their actually having been there. Without being specific, there are a lot of possibilities at Disney for doing that. In Orlando, there's a lot of interesting work going on in virtual reality. I also like Orlando. It's an esthetic place and the beach, and what happens at the Kennedy Space Center is not very far away.

— *How are you adjusting to your new status as an ex-astronaut?*

Who's "ex?"

— *You are.*

I'm not sure how it is. Do I miss it? I've got to. But I can go into space any time I want to.

— *Through meditation?*

Well, through imagination. I put myself there. I do that. Listen, I got dealt a card, and went forward with it.
October 20, 1998

Postscript

We had planned on doing our interview at Disneyland, near Kissimmee where Story Musgrave had moved after his retirement from NASA. I thought it might be fun to do some of the interview while on space-simulating rides. But the September day that I arrived in Florida was damp and hot, and we ended up talking at the grim Orlando hotel that I had booked myself into. "I'm not going to talk about John Glenn," Musgrave declared before I turned the tape recorder on—the Senator from Ohio was about to take his Golden Age space flight redux.

And then, he talked lots about John Glenn. Big surprise. John Glenn at that moment was getting lots of ink as the national senior in space.

Two and a half years after our interview, we talked again on the phone. Musgrave was still working for Disney, as an "Imagineer," which in Disney-speak is probably a kind of consultant.

FREEMAN J. DYSON

A-Bombs, Space Chickens, and God

AP/Wirephoto

In the small world of literary science writing, there are few authors as prolific and original as **Freeman J. Dyson,** 76, the British-born theoretical physicist, astrophysicist, and author of books on scientific matters.

Like Stephen Jay Gould and the late Carl Sagan, Dr. Dyson is a highly credentialed scientist: he has been a professor of physics at the Institute for Advanced Study in Princeton, New Jersey, for 47 years. He enjoys a huge readership, with seven books, including his meditation on the new technologies, *The Sun, the Genome and the Internet.*

One of his marks as a writer has been his willingness to engage in speculative writing that, at times, seems to border on science fiction. It was Dr. Dyson who suggested sending microbes and spores into space in a kind of colonization effort to make the universe a "more interesting place." And it was he who imagined "space chickens," high-tech poultry with the genetically engineered capacity to wing their way through the cosmos.

Dr. Dyson has long used his expertise to investigate how science affects issues of war and peace. His 1984 bestseller, *Weapons and Hope,* discussed the rationale of nuclear weaponry. It was partly his work in this area that won him this year's $948,000 Templeton Prize "for outstanding originality in advancing the world's understanding of God or spirituality."

Dr. Dyson answered questions in his cluttered office several weeks after receiving the award.

___ *The play* Copenhagen *focuses on a visit Dr. Werner Heisenberg, head of the Nazi atomic bomb program, made to his former mentor, Niels Bohr, in Denmark in 1941. In the play, as in reality, the two men took a walk in the park and had a conversation that ended in anger. What actually took place during that stroll is the Rashomon-like puzzle at the heart of the play. Did Heisenberg try to pick Bohr's brain on scientific problems? Was he trying to manipulate his former professor into encouraging the Allies not to build an atomic bomb? Or was he trying to pass on information, or even disinformation,*

21

to the Allies through Dr. Bohr? Freeman Dyson, you knew these men—what do you think really happened in Copenhagen?

I have my own theory, which may, of course, be totally wrong. I think Heisenberg was trying to get Bohr to agree that the development of an A-bomb was something that the physicists of the world should agree to not participate in. Heisenberg was in a position where he could quietly decide "this doesn't make sense." And Bohr had once been in a strong position, too. But by 1941, it was much too late. By that time, they were enemies.

There was a moment earlier when something like that might have had a chance. In January of 1939, there was an international meeting when fission was first proclaimed and when the idea of building atomic bombs was in the air. Scientists could have gotten together then and decided, "This is something that should not be done." Before the war started, nobody really wanted to create an A-bomb. What pushed them, of course, was fear, once the war started.

___ *After Germany's defeat, Werner Heisenberg hinted he had deliberately failed to build a nuclear weapon. Do you think that's true?*

To a great extent, it is. Heisenberg never claimed to have obstructed anything. But he didn't give the project the kind of push it needed.

In countries where bombs were made, there always was at least one scientist who really made a push. Heisenberg was the one who might have done that in Germany. Of course, his not pushing was convenient because Hitler wasn't all that interested anyway. If Hitler had been interested, Heisenberg's position might have been different.

___ *Your latest foray into science policy involves opposition to the newest version of the Strategic Defense Initiative. What is the core of your critique?*

I oppose it because the system doesn't work, mostly because it can be easily outwitted. A very primitive spoofing will take care of it: just not letting you know where you have to shoot. If you know where a target is, you can kill it. It's not easy, but in principle, you can do that. But if you don't know where it is, it's hopeless.

So all the attacker has to do is to make sure that you don't know where the target is, and that's rather easy. All he has to do is hide the missile among a bunch of balloons. That solves the problem from the point of view of the attacker. It is easy to use other kinds of decoys as well as balloons.

— *You've long advocated travel to Mars. Have you been discouraged by the many mishaps NASA's Mars program has suffered?*

No, not at all. I'm a great fan of Jet Propulsion Laboratory. I think it's wonderful they've begun taking risks. They launched six space missions in six months, which is absolutely wonderful. And of course, if you launch at that speed, you take risks and you risk failure. I thought they ought to be losing one out of three. If they were running the thing the way I'd like them to, they ought to lose two or three missions every year probably. To me, that's a sign that they are doing things right.

— *Why are you highly critical of the space shuttle program?*

I see it as a real tragedy. It was supposed to be a compromise between the requirements of getting stuff into space cheap and human adventure. But doing that is like trying to design a railroad where you have passenger and freight traveling on the same train.

We should have something much more flexible for the humans, something more like a sports car than a Greyhound bus. It should be a high performance spacecraft so that the people could actually go somewhere. You could go to the moon, learn how to live on the moon. I think it would be good television, as well. It would be popular. The shuttle is a bore. The shuttle is not going anywhere. It's stuck in one low orbit and that's all it can do.

— *A few years back, you published a proposal in* The Atlantic Monthly *suggesting that future human missions to Mars might carry "warm-blooded plants" which would be genetically engineered to grow their own greenhouses. "Warm-blooded" plants?*

If you want to have a settlement on Mars, you'd better have plants and animals that can live there.

That seems to be the key to any sort of practical colony: plants that grow their own greenhouses. It's like a turtle growing its own shell. You sow the seeds on Mars. You have to pick a place where there is water underground. They put their roots down. Then, they build a greenhouse and they maintain a greenhouse independently. Of course, we are many years away from that. The science isn't in place yet.

— *What was your take earlier this summer, when you saw the announcement of the human genome mapping?*

The genome statement was mostly hype, but that is not important. The important thing is that we now have the tools to sequence all kinds of animals and plants and microbes—as well as humans. It is not important that we didn't actually finish the human sequence yet.

— *The two big revolutions of the past 20 years have occurred in genetics and computer technology. You are as much a futurist as a scientist. Did you see either of them coming?*

Well, the genetic revolution, that was obvious as soon as one saw the double helix. But the Internet, I certainly never heard of it before it happened. It still surprises me that it has taken over the world much more than we expected. It used to be just this little Arpanet that my friends were on and that connected various military bases with each other. We never thought of it as something that would take over.

— *Do you ever feel badly that you were left out of the Nobel Prize given to Richard Feynman, Sin-itiro Tomonaga, and Julian Schwinger for work that all of you did on electromagnetism in the quantum world?*

I don't feel that in the least. All I did then was tidy up work that other people had done. It was fun to do, and that was great. The other thing is, I've been so lucky. I've had this tenured job here at the Institute for Advanced Study, and I couldn't have asked for anything more. I have the freedom to do what I want . . . bright people to talk to every day.

— *Yet, despite all the genius in situ here at the institute, one senses you feel a real affinity for the practical people of science, engineers and the like. Is that true?*

I like people who are working on practical things and who are working in teams. It's not so important to get the glory. It's much more important to get something that works. It's a better way to live. If you are trying to solve the big problems by pure thought, it's very solitary and it's very competitive and the people who don't succeed are mostly unhappy.

— *You've lived a life in science and you have been quite happy?*

Yes. Because I've jumped around many times, and I've never become obsessed with any particular thing.

The other point goes back to what we were discussing about the generation that grew up in the shadow of World War I: My generation took such a tragic view of life. We started out with this deep conviction that life was tragic and that one better make the best of it. One didn't expect anything to come out right. If you start out with a tragic view of life, then anything since is just a bonus.

We did, after all, survive, and we created a world that is moderately peaceful. I take it as a bonus that we only have about 8 countries with nuclear weapons at the end of the twentieth century, instead of the 50 that we expected. Another bonus is that nobody has used a nuclear bomb in anger since Nagasaki.

August 1, 2000

Postscript

In an e-mail in February of 2001, Dr. Dyson reported from his offices in Princeton that, "the main thing I have been doing recently is hobnobbing with the rich and powerful at the World Economic Forum in Davos (Switzerland). Mostly, I was involved in debates, speaking with the voice of reason and moderation against various people with extreme views. I debated Bill Joy about the dangers of biological and nanotechnological terrorism, Brian Greene about the danger that science will run out of interesting problems to solve, Jeremy Rifkin about the dangers of growing and eating genetically modified food.

"While flying back and forth across the ocean, I amused myself with calculations about a little problem in atomic physics that is of no interest to anyone but myself."

"The meetings in Davos were delightful, like a camping trip with a group of exceptionally bright companions, but I do not claim that we solved any of the world's big problems."

ROBERT ZUBRIN

A New Frontier Aboard the
Mars Direct

Brian Payne

Robert Zubrin, 47, an astronautical engineer and founder of the Mars Society, a group advocating travel there, was having coffee in Manhattan on a recent morning.

Dr. Zubrin, who heads the Pioneer Astronautics engineering company, which develops ideas for NASA, had come to New York from his home in suburban Denver to talk about his latest book, *Entering Space: Creating a Space-Faring Civilization* (Tarcher-Putnam).

His 1996 volume, *The Case for Mars,* had made a sensation in astronautical circles with an interesting notion for a cut-rate program for Mars exploration known as the Mars Direct Plan.

The proposed program would use existing technology and would keep fuel weight at a minimum by having the astronauts manufacture propellant for the return trip from material on Mars. *Entering Space* elaborates on the concept.

___ *In both your books, you suggest that the United States will enjoy a kind of national rebirth once it sets up a Mars colonization program. Why do we need a national rebirth?*

I think we need a challenge. Civilizations are like people. They grow when they are challenged and stagnate when they are not. I think this is especially important to youth, who need to go somewhere where no one has gone before. A "human to Mars" program would be an invitation to adventure to every kid in the country: Learn your science, develop your mind and you can be a part of pioneering this new world. In concrete terms, we would get millions of scientists, inventors, doctors and medical researchers. That's the real benefit we got from *Apollo*.

___ *In reading* Entering Space, *I couldn't help but feel you were overromanticizing the nature of nineteeth century America. Yes, it was an era of expansion, but it was also a time of incredible violence, of destruction of the American wilderness and the slaughter of American Indians.*

27

Well, I think people feel the need for a continued existence of an open frontier. They want a place where they can go where the rules haven't been written yet. The most profound kind of freedom that people can have is the freedom to be makers of their own world and not simply inhabitants of it. And that was a sort of freedom that people had during the creation of this country. If we open space, then people can have that freedom again, except there are no Indians on Mars.

— *If there are life forms like microbes on Mars, do we have a right to interfere with their environment? You advocate all kinds of radical activities for your space colonists, like melting the Martian permafrost and mining the planet for minerals.*

If there are microbes, even if we hydrogen-bombed the whole planet to our maximum capability, we could not bomb those microbes to extinction. What we are talking about here is bringing a dead planet to life. We're talking about taking a planetary surface which is, in fact, dead—if there's life on Mars, it's only in the ground water—and turning this into a viable arena for the development of life.

— *In your books, you offer suggestions for the social life of a Mars colony. You have a list of possible rights. For instance, the right to bear arms. Why will Mars colonists need that?*

These are rights that exist in contemporary American society.

— *You suggest that Mars colonists will enjoy the right to bear children and the right to be free of extortionist lawsuits. Where did you get those?*

To be frank, I just sat down here and banged out a whole bunch of rights. In the United States, by the way, we do have the right to have children, but in certain other societies, China, they don't. The point I'm trying to get at here, is what I'm advocating here is not a particular system of rights. These are more a list of suggestions.

— *When you were an engineering student at the University of Washington in 1983, did you ever think your studies would take you into the nether world of politics?*

Not to this extent. Although, I've always been aware that engineering projects had a significant political component. I studied nuclear engineering for a while, which basically failed on the political battleground. The industry has been aborted due to its inadequate performance on the political battleground. It was unable to make its case adequately to the public.

— *Did nuclear engineering face problems because of public relations failure or did Three Mile Island and Chernobyl frighten the public?*

Three Mile Island, of course, was the failure of an American commercial nuclear power plant. No one was hurt. We have much more disastrous things in coal mines all the time in terms of loss of life. Chernobyl wasn't an engineering failure of the American nuclear industry.

But, to return to your question, I did not imagine that I would be leading a crusade. I imagined I'd be participating in political dialogue. I think an engineer would have to put his head in the sand to think that a project could be built without that. The Brooklyn Bridge wouldn't have been built without a massive fight.

— *How large is the Mars Society?*

Right now, we have 2000 dues-paying members and 8000 on our mailing list. We get about 50,000 hits on our Web site daily. Also, we've recruited to our ranks some pretty heavy-hitting people.

We've also launched this project for an research station on Devon Island, in the Canadian Arctic. Devon Island is a polar desert with a meteorite impact crater that has a geology very similar to Mars. You can explore Mars there by analogy.

We're going to build a human Mars exploration simulation base on Devon Island and use it as both an operational and an engineering test base to learn how to conduct human exploration on Mars. We have the money required to build the first module from a lot of small donations and a major one from Steve Kirsch, the chairman of Infoseek. He gave $100,000. At Devon Island, we're going to discover a lot of things that just don't occur to you when you are designing missions on paper.

— *What is your basic take on the recent* Mars Climate Orbiter *mishap?*

First of all, I think the Mars robotic program is underfunded. We used to send our robotic spacecraft out in pairs because they are high-risk.

Now, to save money, we send only one out, and the recent mishap has set the science back four years. The cost of losing the *Mars Climate Orbiter* is low—about 50 cents per American citizen, but the cost in time has been great. So I feel that's why we should have sent two.

— *The Mars mishap was attributed to human error. Is that right?*

I don't know who made what mistake when. But they have reduced the mission operation staff from the previous standard of about 200 to close to 80, as a cost-saving measure and I think they may have gone a bit too far.

— *In the new book, you're rather critical of the nature of aerospace contracts. You suggest that the basic structure of the contracts leads to big clunky programs that take forever to be completed and impede space exploration.*

I think the free market would work better. I think it would be better if the government said to Lockheed Martin, "We'll pay you $200 million per launch instead of 10 percent over your documented costs. If you can do it for $30 million, you can take the rest as profit."

— *When you were a kid, did you have science heroes?*

Einstein was kind of an icon. I actually had an engineering hero as a kid: Admiral Rickover. He said something I tried to take to heart: "The most important quality that an engineer needs to have is courage." By that, he didn't mean physical courage, but the courage of one's convictions. That made an impression.
November 2, 1999

Postscript

Robert Zubrin spent his summer on Devon Island—a polar desert along Canada's most northern frontier. He sees his time on Devon as one big step for mankind toward that ultimate journey to Mars.

Between all that, he has just completed his very first novel, *First Landing,* which Putnam will be publishing in the Fall of 2001.

EARTH PEOPLE

TANYA ATWATER

She Put the San Andreas Fault in Its Place

Michael Tweed

On a warm afternoon recently, **Tanya Atwater,** professor of geology at the University of California at Santa Barbara, one of the few women to be accorded membership in the National Academy of Sciences, sat at her home computer in Montecito, California, making an animated film to explain how earthquakes are created along the San Andreas Fault in California.

Using her own drawings, Dr. Atwater, 57, was producing a film for children that compressed millions of years of geologic history into a few seconds of movement and showed how segments of the earth's crust pushed against one another as parts of California made their way toward Alaska. "Isn't it a miracle that computers can do these kinds of things now?" Dr. Atwater said, laughing warmly. "It's a revolution."

In her own time, Tanya Atwater has seen her share of revolutions. During the 1960s, while still a graduate student at the University of Southern California, Dr. Atwater helped show how the western United States developed.

A paper she wrote in 1970 on the origins and growth of the San Andreas Fault made her a major player in the revolution that was sweeping the field of geology: the realization that the earth's crust is broken into fragments called tectonic plates whose jostling and rubbing together are constantly reshaping the planet.

___ *Is it true that whenever you hear the rumble of an earthquake, you get down on the floor to feel it better?*

Absolutely. When their beds starts moving around, others jump up, but I lay out real flat to feel it. We study earthquakes looking at paper copy of a seismographic record of a quake. That's an intellectual exercise. But if you can really feel the ground, how it's moving, that's something.

___ *Is there a way to paraphrase Dr. Strangelove—learn to stop worrying, and love the quake?*

All the things we find beautiful in landscapes—mountains, cliffs—wouldn't exist, if it weren't for tectonic processes, like earthquakes.

— *How close is your home to a fault?*

There might be one under the whole coast here. That's a big debate. We are probably in a situation like Northridge; this is a subsidiary fault, not the main San Andreas. So we're probably only going to have earthquakes every 1000 years. In my lifetime, is it going to go? I don't know. With the San Andreas, we know it goes every 100 to 150. Our whole section hasn't moved since 1857, so we are right in the window. That's why Californians make such a big deal about The Big One.

— *What should Californians know about The Big One?*

That we live on a plate boundary. The rim of California is being hauled off to Alaska as we speak. It isn't being hauled off steadily. Every now and then, it rips and moves another few meters. The earthquakes are really just showing us that the plates are going by each other. You're not going to stop the whole Pacific plate from moving toward Alaska.

Now, when we're talking about The Big One, what geologists do is look at the faults that make up the plate boundaries and ask if they've moved lately and whether that motion between the two big plates has been picking up or not. So, there are stresses in the fault that have not broken in a very long time. For instance, the one that wraps all around Los Angeles and is kind of inland from here, hasn't broken since 1857.

— *So you would not buy a house in San Bernardino, Riverside, that neighborhood, because of the faults?*

I'd buy a house anywhere, if it was built well.

You know, earthquakes, they are like weather, and we aren't afraid of the rain. We make good roofs, good gutters and good drains. Similarly, we know how to build buildings that don't fall down in earthquakes. The truth is frame buildings don't fall down. Their chimneys do, sometimes. Actually, buying a house in an earthquake zone is much less scary than buying a house on a beach cliff. Now, people who do that are crazy!

— *How did you become a geologist?*

I started out as an undergraduate in physics at M.I.T. in the early 60s. My mother was a botanist and my father, an engineer.

So I grew up in a scientific family. At M.I.T., I took geology and I just loved it. As a kid, my family traveled around the West, camping, river rafting, exploring. Finding a science that got me outdoors and explained things I loved was irresistible.

However, geology in those days was a lot of facts without a central template explaining why things were the way they were. Happily, just at the time I was getting into it, the plate tectonics revolution was starting with all this new data from the middle of the ocean floor that showed how the plates moved apart.

So eventually I applied to graduate school at Scripps Institute of Oceanography. I got there two weeks after Fred Vine, who was the daddy of this theory about sea floors' spreading. He had all these magnetic measurements of the ocean floor. It turned out that the ocean floor has markings on it, magnetic stripes, that with a magnetometer show you how old it is. The data that you get is so clear and easy to read that it just changed everybody's idea about everything.

— *How did you get from the sea floor to California?*

The pattern of stripes off the coast of California was different from that of other oceans. We concluded that there were missing stripes, which had slipped under the rim of the North American continent, bringing the Pacific plate up against the edge of the continent where it could rub and create the San Andreas Fault. What I did was explain the San Andreas Fault in the context of the larger global scheme—something that hadn't been done before.

— *You had big success very young. How does a successful scientist top herself?*

Living up to my old self has been very tough. The early success, it's given me great power: the ability to choose whatever I want to do. I don't have to make a reputation and this has given me the choices that a lot of other people don't have. For instance, I spend a lot of time working with K through 12 teachers. I teach undergraduate geology.

— *Geology wasn't a field for women when you started. What obstacles did you run into?*

Blatant sexism. A lot of institutions didn't allow women onto research ships because they were considered unlucky. They wouldn't allow women on off-shore rigs. There were no women in Antarctica; I applied for that.

Luckily, I had male mentors who helped me break barriers. When there were administrators who said I couldn't be on a small ship because there wasn't a separate bathroom for women, my mentors would say, "Well, she could sue you." And suddenly, they discovered they could manage to have a female on board if you locked the door on the bathroom. For a long time, women were not allowed in the submarine, and I needed to go down to the ocean floor. Their main reason was, "how is she going to urinate?" I was there doing spectacular science and what they were worried about was that!

—— *In John McPhee's book* Annals of the Former World, *he claims that most geologists go crazy at the sight of a good road cut, a place where parts of a mountain have been sliced through to make way for a road. Do you?*

Oh, yes, I almost crash the car when I see a good one. Around southern California, since the San Andreas is sort of ripping through the countryside, the whole rim of California is all just folded up and the layers that started out as flat sediment are all contorted and set on their ears. In a lot of road cuts, you can see that.

—— *Have the changes in technology—high-speed computers, satellites— changed the way geologists work?*

Oh, yeah, particularly the information we get from NASA. For decades, the geologist's job was to go on a ship, go out at 10 miles an hour and run little lines across the ocean to collect magnetic and photographic data. That's still true for magnetic data. But the stuff NASA is collecting from up above is just a miracle. Now, a satellite looking at the top of the water can detect the topography of the ocean floor. The satellite goes around the earth every four hours and produces maps with incredible detail. It does a much better job than we'll ever do.

—— *Geologists deal with tens of millions of years at a glance. What do you feel when you're trying to understand events over such a vast period of time?*

Very tiny and ephemeral, really humble.
October 12, 1999

Postscript

This was received from Tanya Atwater on March 18, 2001:

"Can't give you an update. Most of the things we talked about were lifetime passions, not things that change much in a year or two."

"My job title is "Full Professor.""

"My official spoken title is "Professor," beyond that, it is whatever I say . . ."

"I guess I'd say 'Professor of Tectonics.'" ("If I were in Russia it would be 'Academician'—as in 'Academician Tatiana Evgenovna Atwater')."

JOHN McPHEE

A Writer Takes a Turn Reading
Sermons in Stone

Laura Pedrick

John McPhee, 67, a *New Yorker* writer best known for taking long looks at small things, has recently published *Annals of the Former World,* a 696-page look at the geologic history of a large thing: North America.

In that work, which took Mr. McPhee much of the last 20 years to produce, he and other geologists travel from the Newark Basin to the San Andreas Fault as they uncover the prehistory of the land.

Mr. McPhee spoke with *Science Times* recently from his offices at Princeton University, where he is a professor of nonfiction writing. He said he devoted himself to the *Annals* because he found geologists and what they do a "completely engrossing subject."

—— *How did a 20-year book project on geology change your perception of the world?*

I came to look at rocks literally, specifically. For instance, looking at an outcrop by a road: It's not what it appears to be. It's not what it used to look like. Its history is embedded in its rocks.

Another thing that changes on a project like this is your sense of time. We live in such a whisper of time. Developing a sense of geologic time is frightening. Initially, you come up against numbers that are so large that they don't really mean a lot to you—even if they differ by millions of years. But once you get a sense of that, when you realize that one million years is really a short unit in the history of the earth, that the earth is marching along and constantly changing in its own time scale, you come to feel absolutely insignificant. The other side of that is that you feel a little less disturbed about the human condition. I think it helps in your contemplation of death.

—— *Science writers often have trouble making the vastness of time comprehensible to readers. How did you do it?*

You use analogies, a lot. Also, if you just live with this problem long enough, you come to see it. You realize that the time of the dinosaurs is very, very recent, comparatively. It's only 65 million years ago. The earth is 4.6 billion years old. Once you've grasped that, you can close your eyes and really sense the motions of collision and mountain-building and oceans appearing and disappearing. You can see it going by, in your mind.

— *Give us a geologic description of the place we're talking from this moment: East Pyne Hall, in Princeton, New Jersey.*

Well, Princeton is in something called the Newark Basin. Two hundred million years ago, there was country here that was part of a super continent and got active and started to pull apart. The earth's crust pulled apart here and broke into blocks separated by basins, and the largest separation is now 3000 miles wide: the Atlantic Ocean.

All up and down eastern America are places where the earth pulled apart and basins formed and sediments poured into them from all sides. The Newark Basin is one of them. The rock surrounding the Newark Basin is 400 million years old, or older. This rock here in Princeton is only 150 million years old because it's the sediment that fell into the basin. Most of it fell into a great big lake that formed here in the basin and lasted eight million years.

Nassau Hall, the building next door, is made of Stockton sandstone, which came from down the hill by the lake a mile away and was quarried in 1756 or something like that—very local rock. It was stuff that fell into that lake that I've been describing. The library is Wissahickon schist and it doesn't come from here. It was lugged in trucks from Pennsylvania.

— *When you walk around town, or drive on a road, do you now deconstruct the composition of every building you see?*

Do I do that? Yeah. Ask my wife. She says she'll allow me to do that 10 minutes a day.

— *Tell us about geologists—you hung out for two decades and traveled from coast to coast with them—is there such a thing as a geology-type?*

Well, beyond the fact that they tend to carry hammers, wear railroad engineers' hats and wool shirts and jeans and scuffed boots and carry Brunton compasses, no.

One generalization I will make is that geology often enough attracts people who originally were trained in another science. They seem to get drawn to it because geology is eclectic. Geology involves physics, biology, chemistry. Jason Morgan, one of the signal people in this or any other science, got his Ph.D. in physics. He's one of the cornerstones of plate tectonics. Ken Deffeyes, who I did much traveling with, was a chemist and then became a geochemist.

— *One gets the sense that within the academy, geology is looked at as a slightly inferior science because it is not so theoretical—it is, in fact, quite earthbound.*

Well, that could be, although that aspect is arresting. On the other hand, geologists developed plate theory in the 1960s. In a period of eight years, a series of papers were published that absolutely revolutionized our understanding of the earth—completely. Before that, nobody knew that mountains were built by intercontinental collisions, that oceans were formed by extensional disassembling of a piece of the earth, which eventually opens up like the Red Sea. Anyway, this stuff was unknown before these people performed the scientific revolution.

But, supporting your theory, the number of geology majors in many a university's undergraduate population probably will be influenced by the price of oil. The fact that this subject is so interesting per se does not cause universities to roll over and increase their budgets for geology departments. Academic geology is very much related to the world of mining and oil and so forth.

— *You seem to enjoy geology humor. Do geologists have a witty side?*

Geologists are fond of puns. And it doesn't seem to be very conscious. I've pointed this out to some of them, and they scratch their heads. For example, one of their primary journals is called *G.S.A. Today.*

— *Why do they refer to the Pocono Mountains as "the so-called Pocono Mountains?"*

They don't. *I do.* In my book. Because they aren't mountains. The Poconos are a flat-lying plateau that's dissected by streams. The Catskill story is the same.

— Is it difficult to develop narratives about rocks?

I find it very difficult. Well, I find it very difficult to write about anything. It's particularly hard when you are trying to grasp something that is virtually in another language and then try to describe it in a way that will make sense to someone else not trained in the field.

— Is it true that in your past you'd tie yourself to a chair in order to write?

Once. In the 1950s. I had a captain's chair with spindles and I sat in there by a typewriter and tried to write. One day, I took the sash from a bathrobe and tied it between the spindles and then around my waist. I did it three or four times. I abandoned that. It didn't work.

— Back to geology, do you think most modern city-dwellers underestimate nature?

Oh yeah. The seashore is full of people trying to stop nature in her tracks, and she doesn't want to stop. Coney Island wants to go south. Hatteras Light is sitting on ground that wants to move. But we feel it's a National Historic Site—so we're spending millions of dollars to fight a battle that probably is not winnable.

— When you think about all this, does it make you religious?

I'm really impressed. This subject that I have worked on for 20 years is nothing more or less than describing the earth and the people who study it. And why am I doing that? Well, this is the only earth I'm ever going to live on, and I find it an interesting place.
November 17, 1998

Postscript

Not long after our chat, John McPhee won a well-deserved Pulitizer Prize for _Annals of the Former World_. In March of 2001, I heard this from him: "Dear Claudia: I'm just back from Spain. I begin another trip Monday so forgive me if I'm brief. For a couple of years I have been working on a book on American shad (Alosa sapidissima)—their natural history, their American history, and my own experiences with them as a shad fisherman. Several chapters have appeared in *The New Yorker*. I hope to finish the whole thing in 2001. I'm still teaching my Princeton University writing course.

"And about the 'books within Geology,' I consulted with the tectonicist Eldridge Moores at the University of California, Davis, and here are some titles: *Understanding Earth,* third edition, by Press and Siever, an introductory text published by W. H. Freeman and Company, New York, 2001. *Dr. Art's Guide to Planet Earth for Earthlings ages 12 to 120,* by Art Sussman, published by WestEd, www.wested.org. *Earth Shock: Hurricanes, Volcanoes, Earthquakes, Tornadoes, and other forces of Nature,* by A. Robinson, London, Thames & Hudson. *History of Life,* second edition, by Richard Cowen (Blackwell, New York).

"All best, Claudia. Hope this helps. John"

MICHAEL J. BALICK

New York's a Jungle, and One Scientist Doesn't Mind

James Estrin/*The New York Times*

At 47, **Michael J. Balick,** who directs research and training at the New York Botanical Garden, is among the few scientists in the world to earn a living as an "ethnobotanist."

"Among the things I do is wander through New York City looking at urban markets, asking healers in botanicas, 'What do you use this for?'" Dr. Balick, the co-author with Paul Alan Cox of *Plants, People, and Culture* (Scientific American Library, 1996), explained on a recent morning in his office at the gardens in the Bronx.

"I also advise physicians on the methods of traditional healers," he said. "What can be more interesting, or more fun?"

___ *How did ethnobotany become your life's passion?*

I was one of those kids always taking leaves from the woods and pressing them between wax papers and ironing them. Forty years later, I'm being paid to still do that.

___ *Give us the job description of an ethnobotanist.*

Someone who studies the relationship between plants and people. That means the foods that people eat, the spices in their foods, the medicines they use, and how people manage their resources, forests, farm lands.

For the longest time, this profession was traveling to a distant culture and collecting plants. In the last 20 years, the paradigm's been changing.

What we now do is add the modern tools of science to this old discipline. Also, we educate. We try to help urban peoples see how much we all depend on plants. Plants are the bottom line in life. They support all of life.

___ *Do you have a favorite ethnobotany movie?*

Medicine Man. It featured a couple of big ethnobotanist fantasies in it. The Sean Connery character, the ethnobotanist, had a high pressure liquid chromatograph, which is a very sophisticated analytical device and he had it set up in a remote indigenous house—with no electricity!

That's always been my dream. To have the latest scientific equipment down in the middle of nowhere.

— *What kind of equipment do you actually take to the field?*

Very simple tools: a machete, a still camera, a video camera, a backpack, a hat, sun screen, a plant press, some old newspapers.

The second fantasy in that movie was when Sean Connery wrote for a research assistant, Lorraine Bracco. She paddled up in a boat, and the first words out of her mouth were: "Dr. So and So, our company has been sending you money for nine years, and we never have even gotten a report from you once!" That to me was the greatest! We are swamped with report-writing.

The general public thinks ethnobotany is a very romantic activity. But the truth of it is that you're traveling for long periods of time—not with your family—to isolated places. You're living in miserable conditions most of the time, covered with insects.

You're often sitting around waiting for things to happen because you can't force them.

Actually, it is a very difficult profession. I could tell you stories of people who said they wanted to do this work and then when they finally got in the field, they wouldn't come out of their hut.

— *What's your best personal story from the field?*

In 1976, my wife, Daphne, and I canoed up the Ampiyacu River in Peru, looking to spend time with the Bora Indians.

And when we eventually arrived at a village, 40 or 50 people came out and circled us. They had not seen many of our kind. We were a great curiosity.

While we were explaining what we wanted to do, suddenly an old woman came out of the Long House and hugged me. Everyone fell on the ground laughing.

So Daphne asked, "What is grandmother saying?"

And the village chief says, "You don't want to know." To which, Daphne says, "Oh, no. Please tell us."

So he explains, "Grandmother said that what a great opportunity it would have been to have this guy come to our village three years ago before the missionaries came here, when we were still eating people because he could have fed us for a month!"

And she proceeded to describe how you cook the fingers, the hands, the back. Three or four years previously, before the missionaries came, they were practicing cannibalism.

We lived with them for six weeks, and they were just the most hospitable and giving of people.

They showed all these uses of plants: how they grew things, how they cultivated their tubers, how they harvested things from the forest, how they made backpacks from palm leaves.

— *I understand you do a lot of fieldwork in a different kind of forest—New York City.*

Oh, yes. One of the most diverse laboratories that an ethnobotanist could have is New York City. All these different cultures can be found here.

They all bring their healing plants, food plants, relaxing plants, beverages with them. I give a class at the Yale School of Forestry and Environmental Studies. I do my shopping for it in Chinatown and Harlem and also Dean & DeLuca in SoHo. They have amazing tropical fruits that no one else has.

— *Do your two worlds—New York and tropical rain forest—sometimes clash?*

Rarely. Though once, I was home in Westchester unpacking my collection of rain forest hunting tools. And I had this dart set that had curare poison on its tip.

I was very tired and I stuck myself and I started feeling a little funny. So I called up the number of Poison Control and they said, "You should call Dr. Balick—he's our specialist." And I said, "Man, this *is* Dr. Balick!"

So the woman on the other end of the phone asked my symptoms and said, "Now you get yourself to a hospital, but you call in advance first."

I then called one of the local hospitals and said, "I have a curare stick wound from a poisoned dart, and I'm coming in an ambulance." And the physician in charge, thinking it was a joke, hung up.

I eventually ended up in the E.R. of another hospital—where there were people with heart attacks and knifings; I was the guy in the corner with the poisoned dart wound from the Amazon.

___ *The use of botanicals and herbs in western society has increased. Do you find this a positive trend?*

Well, it is fascinating. In the last few years, probably over a third of the American public has begun using botanical medicines for some aspect of their health care.

People now know if they get into a car wreck, they go to an emergency room, but if they want to stave off a cold, they might take echinacea. If they want to help with their memory retention, they might take gingko.

Actually, I only recommend that people utilize herbs under the care of a trained health care professional. There are specific situations where you shouldn't be using herbs. For example, if you have a disease of the immune system like lupus or scleroderma, you shouldn't be using echinacea.

I'm also concerned about the growing use of botanicals in the personal care industry. I'm worried that in some cases, this use—without any thought of sustainability—could lead to the demise of some plant resources.

Many manufacturers have jumped on the bandwagon of sustainability these days, but in truth, they don't have a clear idea of what that means.

Sustainable harvest involves removing only that quantity of a plant or a plant population that will insure that the wild resource continues to thrive.

Not all personal care product manufacturers who say they "help the environment" actually do that. I had a student look at some of these manufacturers who claim, "We're giving back to the rain forest."

She called some of the companies that bragged on their labels about how much they were giving back to the environment. It turned out that one of the bigger ones was giving away $140 that year and another one said, "We had no profits so we didn't give away anything."

___ *In all your years in this business, what would you say has been your greatest accomplishment?*

Working with people in Belize—Dr. Rosita Arvigo and Gregory Schropshire, and a community of traditional healers—to help them preserve an entire forest, an entire ecosystem. And we call this, the world's first ethno-biomedical forest preserve.

It's a forest preserved so that traditional healers would have a place to sit, teach students, plant things that are being destroyed in other places, and gather the materials for their art.

The other thing I do, which I love, is work with urban young people who come to the Botanical Garden from the neighborhoods of New York. If you watch kids, they have boundless enthusiasm for nature. They will make observations adults never would, like: "This flower smells just like old socks!"

You want to maintain that level of enthusiasm and openness to nature. Indeed, I think people who really enjoy this kind of work have maintained that persona of a child's awareness of the earth and humility to it.

— *Do you keep plants at home?*

Some. I keep plants that can go without care or watering for a very long time. Because, obviously, I travel.

I keep dracaena and yuccas and other things like that. I have an herb garden and I struggle with it. My house happens to be situated on a shaded plot.

I grow tomatoes, and they will cost about $40 a pound, if you calculate what I've spent on them at the end of the year. But it's money well spent. The therapy is in the growing.

April 6, 1999

Postscript

Though this was hardly cause and effect, shortly after our interview Dr. Michael Balick was given the weighty title of "Vice President for Botanical Science Research and Training Philecology Curator and Director, Institute of Economic Botany, the New York Botanical Garden"—known to most New Yorkers as the "Bronx Botanical Garden."

It is common for modern academics to assume three-thousand pound titles. And, in my opinion, it's an awful practice that often leads to lethal solemnity. Michael Balick, when we met in the spring of 1999, was a funny and inventive charmer. He had that quality that many the best scientists (and artists) have, which is a childlike playfulness.

But whoaaa!!! What was this???? When, in February of 2001, I e-mailed him to get an update on his life, what I got back was a request that we cut the previously published anecdote about a tribe of former cannibals he'd once visited with because, apparently, there were some scientific friends who didn't like him telling that story.

I wrote back: "Can't modify it . . . certainly can't because of issues of scientific and political correctness. But meanwhile, I'd be grateful to learn of your most recent travels and general fun activities in the world of ethnobotany."

So the next thing that Dr. Balick writes me is that he had an E-coli infection, contracted on a visit to China, and that while he was in Micronesia, he feasted on a dog.

All I could do was hug my Cairn terrier and write back, "What breed?"

Dr. Balick didn't know. And this guy thinks he has trouble with colleagues who think it uncool for him to repeat a story about a tribe of former cannibals looking at him like luncheon meat. Dining on dog—whatever the breed—is, in some circles, the ultimate in incorrectness.

MARTIN WELLS

He Studied Squid and Octopus, Then He Ate Them

Jonathan Player

Martin Wells, 70, grandson of H. G. Wells, retired reader in zoology at Cambridge University, one of the world's great experts on cephalopods and author of the just released *Civilization and the Limpet* (Perseus), can speak for hours about the wonders of cuttlefish, squids, and octopuses—his specialty and great passion. Dr. Wells held forth on a recent November afternoon in his sprawling seventeeth-century house in Cambridge, England, while his wife, Joyce, put together a supper of steamed vegetables and cheese.

— *If there were such a thing as a vivisectionist's Anti-Defamation League, you'd be heading it. What's good about vivisection?*

Knowledge! I am proud to say that I've been a practicing vivisection-ist all my life. I work on octopuses and squids and things like that, when I'm not eating them.

My point is, if you want to conserve animals, you've got to know how they work and what they need to survive. There are a good many things that can only be found out about animals, and people, by doing experiments.

If you look at squids and octopuses—my specialty—practically every darn thing we know about how nerves work, has been worked out from squid nerves. That's because squids have good, big, nerves you can put electrodes into. Unless you're prepared to cut up a few squids to get that material, there's no way we'd learn how nerves work.

— *Did you say you eat your experimental subjects?*

Oh, when my wife and I were first married, I had a job on the staff of Naples Zoological Station in Italy, and we were very impoverished. So of course, we ate our experimental animals. The only thing we had to do for the laboratory was save their brains. We ate calamari till we got pig-sick of it.

— *Why is squid, by and large, your specialty?*

I find them fascinating. The whole group—squids, octopuses, cuttle-fish—are interesting because they are animals, which, by and large, live on their wits. They react to the same sorts of things we react to. You can look at them and understand their behavior. Another thing I like about squids is that they are predominantly visual animals. They see things coming and assess them. They have to learn a lot in their life because they grow from millimeters to meters long, and if you think about that, you realize that who is an enemy and who is potential prey alter throughout the lifetime. And they've got to be adaptive in that way.

— *Do squids have anything like a language?*

It depends on how you define "language." If you define it as anything that has grammar and syntax, then you are in trouble. Some people would say that squids have a form of communication and can indicate quite complicated things like, "Hey, there's a predator coming, chaps—let's move over." Squids can signal to other squids what sex they are and what condition they are in. "Is it worth approaching me with a view to mating or not?" Or, "Get out, I'm bigger than you are, and I'm after that squidess over there." Whether they can say anything more complicated than that is to be debated.

— *Your books on zoological subjects are breezy, easy to read. How do your Cambridge colleagues feel about your being . . . a popularizer?*

I think it's rather approved of, though that wasn't always true. When I first joined my department in the early 1950s, there was the feeling that people who did broadcasts and wrote pop stuff should keep quiet about it. And I used to write BBC pieces that went on in the half-times of concerts. They'd buy anything—so long as it lasted exactly six minutes. So, I wrote pieces on snails and bats, and that raised a few eyebrows.

One of the things I wonder about now, is—why bother to write pop books when there are all those marvelous nature programs on television? You look at some of these and you think, "How much work has gone into getting those shots!"

The stuff that David Attenborough does is certainly wonderful. He was almost a contemporary here at Cambridge. He took the same courses

I did, and in fact, the big *Life on Earth* series, which really put him on the map—I could recognize large chunks of it.

— *You've written that you never wanted to be anything but a zoologist. When did you know for certain?*

As a child, I used to collect things out of the river and put them into aquariums. I went to nearby rivers for specimens because this was during World War II and you couldn't get near the coast—it was mined. So I never met any marine animals at all, until I was sixteen or so. After the war, my parents took me to France and there I discovered a mask and snorkel and saw things that I had only heard about in books. That's what got me into it.

— *David Attenborough once told me that as a boy he also collected butterflies and lizards. Is the practice a sign of a budding zoologist?*

It used to be. The sad thing now is that in my department here in Cambridge there is only a handful of us who probably can tell one animal from another. The rest are molecular biologists and model-makers. There are sort of mutually incomprehensible languages taught in my subject now. My elder son is a veterinarian who has gone into molecular biology to do research into muscular dystrophy. When he and his wife are discussing their work, I practically can't understand them.

In the biology department in Cambridge, you'll meet more people who know about the movement of calcium within cells than who know how a mollusk moves. Now, I do think it's good to move a subject forward. But the sad thing is that they miss so much. Whole animals, whole plants, are such lovely things. I just can't feel the same way about a molecule coursing its way through a cell.

— *Have you read much of your grandfather's science fiction?*

Of course. I think of him as one of the first popular writers to cash in on his scientific education. He got scholarships to come to London and learn biology under T. H. Huxley. But in his time there weren't really any posts for professional scientists. So what H. G. did was use his education in stories. *The Time Machine* was a straight application of the idea of evolution into the future.

— What are your recollections of your grandfather?

When I was a child, H. G. used to come down for the weekends and play with us, the grandchildren. He died when I was about sixteen. The problem was that the time when I might have got to know him was during the war. And he rather bloody-mindedly insisted on living in the middle of London. And, of course, you didn't take the grandchildren up to London during the war because it was being fairly consistently bombed. So apart from the occasional weekend, I never really got to know him. Which is very sad. Because he was a biologist. I would have liked to gossip with him. But by the time I had gotten into the biology fray, he had faded out.

— Have you tried your own hand at science fiction?

Yes. I've done a couple of novels, which I haven't sold to a publisher. I'm a fan of the kind of science fiction where you are allowed about one variable that you can alter and then you follow through the consequences of that. And my own pieces are like that. For instance, I've written a genetic engineering story about triggering elements of a rerun of vertebrate evolution. It starts out with a young man who works on amphibians and who's brought in as a consultant because all these strange amphibians have been turning up. This triggers a whole lot of events which end up with the evolution of primates again.

Another story I've written is about dolphins becoming a plague—they become a ravenous beast in response to overfishing. The only adaptive response that dolphins could make to it was to breed more readily.

— One senses from this that you're not such a fan of everyone's favorite marine mammals—whales and dolphins.

I find dolphins overrated. They are beautiful animals, but I'm not sure they're as smart as they are made out to be. And on what grounds whales are supposed to be intelligent has always rather beaten me. Anything that spends its life looking for rather easy-to-find krill isn't much better than a cow in the field, is it? If what you eat is grass, grass doesn't struggle to escape very much. No, I can't believe that whales are very smart. Maybe killer whales, because hunting takes skill. The most interesting animals are the ones living complicated lives because they have to solve a lot of problems.

I think I'm perhaps unusual in that I find many of the invertebrates as interesting as birds and mammals. Nautilus snails, think about them: They are ancient! It's a mystery of an animal and it lives at great depths. It's one of these classic living fossil animals.

— *So let me ask a Barbara Walters sort of question, if you could be a Nautilus snail or a squid, which would you be?*

Squid, without a question. I think its life is more interesting, but deeply stressful.
December 8, 2000

Postscript

I've run into Martin Wells a couple of times since our interview. He's a charming raconteur—and also a fun correspondent. When I e-mailed him in February of 2001 to hear of his latest adventures, I got back this travelogue:

"No books since *Civilization and the Limpet*, but one screenplay, 'Meercat' (about another species stealing fire and a diamond scam), which is being looked at by film people in South Africa.

"And I'm rehashing a novel called *Dolphin* (about nice animals that become a plague, animal liberation and a vet in the Western Highlands who becomes pregnant by a seal) that I failed to sell in an earlier version a few years ago.

"Otherwise, I paint (animals, boats, girls); I have an exhibition booked for next year. I caught a lot of trout in the summer. Joyce and I spent the late summer–early autumn sailing our boat in Greece, Christmas and the New Year in Los Angeles with Son Two, Simon, wife, and our two grandchildren.

"Simon is in your part of the world this week, shooting winter scenes for *The Time Machine*, which he is directing for Dreamworks/Warner Brothers. Trip to USA only slightly marred by a snowboarder who wiped me out at Mammoth, I was bruised but not broken or notably concussed, a lucky escape. Old idiots of 72 shouldn't ski?—but I now get ski-lifts half price. Now off to the Galapagos, zoological pilgrimage, and Chile, visit cousins.

"Our timing on foreign travel is impeccable, last year we booked to take Son One and family to Zimbabwe, more cousins, expedition cancelled by the awful Mugabe. Yesterday we learn that a state of emergency has been declared in Ecuador to which we fly on Monday. And that there is an oil slick in the Galapagos. Going anyway. And we seem to have coincided with small earthquakes in Greece and L.A. We look forward to earning a modest living taking bribes not to visit unstable countries."

I also received the Wells' family recipe for calamari:

"First catch your squid. They should be Loligo vulgaris of good Mediterranean descent, not more than six months old, and caught in the last twelve hours. You need four or five squid, mantle length four or five inches, per person.

"Cut off the heads and tentacles. Remove the beaks (the hard round dark bits at the front end of the head, inside the ring of tentacles). Remove the 'pens' (the pen is the transparent stiffening rod down the back of the body, you can grab one end in the nape of the neck after cutting off the head and pull it out). Remove any guts that you can pull out of the cone formed by the muscular body.

"Prepare the stuffing, these quantities will do for eight or ten small squid: In a frying pan add one or two chopped cloves of garlic to a little olive oil and cook slowly until the garlic is transparent and starting to brown. Chop up the heads and tentacles. Add to the oil, then add half a dozen stoned and chopped black olives, a tablespoon of capers, a tablespoon of chopped parsley, a pinch of hot red pepper seeds (or a slosh of Tabasco). Brown that lot and then add a tablespoon of chapelure (fine dried breadcrumbs) to dry the stuffing so that it can be handled and pushed into the cones of squid. Close the cone entrance with a toothpick.

"Use butter in the frying pan to brown the stuffed squid. Then add chopped and peeled tomatoes, either fresh or tinned, stiffened with a little tomato paste, enough to almost cover the squid, add a crushed clove of garlic and cook gently until you can stick a fork into the squids.

"Last time I made it I used frozen squid, species unknown, caught in Bombay. They were still very good. Maybe your friends will still speak to you after all that garlic. Best Wishes, Martin."

LUIS F. BAPTISTA

Revolutionary Etude, Rendered
by a Wren

Peter DaSilva

San Francisco

On a drizzly afternoon here, **Luis F. Baptista,** 58, an enthusiastic man with dark hair and the accent of his native Macao, was showing a New York visitor the ins and outs of daily life in this city's Golden Gate Park.

"This is where a white crowned sparrow I used to know once lived," Dr. Baptista said, pointing to an ornamental African aloe tree. "She was very dominant to her husband, whom she beat up on all the time. However, one day, she divorced him, moved to a different tree a few hundred feet away and married another guy who then beat up on her. I couldn't understand what she saw in him. But they raised more white crowned babies than anyone else in the park. She must have realized there was something special about the new guy, in terms of evolution."

Dr. Baptista knows all the soap opera details of the lives of Golden Gate Park's birds. As chairman and curator of ornithology and mammalogy of the California Academy of Sciences, based in the park, he has spent a lifetime studying the mating habits, songs, and language of birds. The author of some 127 papers on the subject, Dr. Baptista is among the world's leading experts on bird song, dialect, and language.

As he sat under the tree of the much abused and abusive white crowned sparrow, Dr. Baptista speculated about an ancient question.

___ *In the 1950s rock classic,* Why Do Fools Fall in Love, *the singer Frankie Lymon, asked "Why do birds sing so gay?" Do you know why?*

They sing, for one thing, because the bachelor male wants to attract a wife. He also wants to show other males that he's pretty studly, that this is his territory, and no one is going to cross the boundaries. A third reason is that the song induces the female bird's brain to send hormones: it makes her ovaries grow.

We can actually show that birds have an aesthetic sense similar to ours. If you take female great reed warblers, stick them with female hormones, put them in a soundproof box and play for them recordings of male songs—some beautiful and some yucky—you will see that they solicit copulation more from the beautiful songs than the yuckie ones. That means that the females actually prefer beautiful songs, even what humans consider beautiful.

— *Are there cases when humans, like certain birds, have a sexual response to songs?*

I suppose there are. You remember all those Beatles concerts and the girls passing out to the sounds of the Beatles and Elvis Presley? For me, it's Montserrat Caballe. When she sings, I feel like I've gone to heaven.

— *Do you think, as it has sometimes been suggested, that birds sing for the pure pleasure of it?*

There is a possibility. Konrad Lorenz (the Austrian zoologist) had a jackdaw, and he said that when the bird would have a nice meal and was relaxed and had no stress, it would sit on a tree and sing. It would use all the calls in its vocabulary and he would do these movements like a Shakespearean actor doing a soliloquy. And Lorenz suggested it was a form of play. So in that sense, I would say that it is quite possible that birds—when they are happy, well fed, without stress, without predators— sing just because they feel good.

— *Haven't you also noted that some birds are instrumentalists?*

Oh, absolutely. The black palm cockatoo actually shakes a piece of stick that it uses as a drumstick. Then, it holds it with one foot and bangs it against a hollow log, as he's calling and displaying to attract the female.

In this country, we have lots of woodpeckers. Woodpeckers peck, of course, to get bugs. But they also peck to make music because they don't have a vocal song like song birds. So what they do is they look for a nice log that makes the correct sound and then, they pound on it with a specific rhythm. This means that tool making is not an exclusive property of humans or other primates.

— *Do birds have language?*

They don't have a grammar. The definition of a language to a linguist includes grammar, but they do use different sounds that communicate different messages. A chicken, if you show it a picture of 'a' raccoon, it will recognize it as a predator and make a certain clucking sounds and all the other chickens will look around for a raccoon. On the other hand, if you show it a picture of a falcon, it makes a different sound and looks up, which means "the danger is in the sky." Song birds have that too.

— *Some scientists insist that the particular sounds that birds make are just instinctive reactions. How would you answer them?*

I guess the same could be said for humans then because we instinctively react to certain kinds of sounds, right? I don't see the difference.

There are some bird sounds produced that are learned. I studied a European bird called the chaffinch and it has a vocabulary of 10, 12 different sounds. But it has different alarm calls. There's one for the hawk flying overhead: "This is not a serious danger." A different call means: "Let's get the hell out of here."

— *You study bird dialects. Why would a bird have a dialect? Is there an evolutionary reason?*

My favorite theory is that the structure of a bird song is determined by what will carry best in its home environment. Let's say, you have one bird that lives in a coniferous forest and another in an oak forest. Since the song is passed down by tradition, then let's say there's an oak woodland dialect and coniferous woodland dialect. If you reproduce the sounds, you will find that the oak sound carries farther in an oak forest than it does in a coniferous forest, and vice versa.

— *How did bird dialect and music become your passion?*

It started when I was a boy in Hong Kong. I went to teahouses where people kept their song birds on the top floor. The practice enchanted me. I also kept two parakeets and then I had canaries. I had young canaries and I was impatient to get them singing. So I started singing to them. To my amazement, when they were adults, they sang just like me.

— *What are the parallels between human and bird music?*

I know of birds who have voices with tonal qualities that sound like real instruments. The strawberry finch has beautiful single notes that come down the scale and that sound just like a flute. There is another bird, the diamond firetail from Australia, whose voice sounds like some kind of woodwind, an oboe perhaps. Then, in Costa Rica, I've encountered a wonderful night bird, and it sings four notes coming down the scale, and the quality of its voice is just like a bassoon.

Then, if you look at pitch, scholars have found that certain birds use the same musical scales of human cultures. One scholar has found that the hermit thrush actually sings in the pentatonic scale used in far eastern music. One of the most incredible cases is the canyon wren, who sings in the chromatic scale, and his song reminds me of the introduction and finale of Chopin's *Revolutionary Etude*.

___ *How do you tell when convergence occurs? And when is man just borrowing sounds heard in nature?*

In this case, it is really a convergence because there's no way that Chopin ever got to where this bird lives: California. However, there are real cases of borrowing. In the Sixth Symphony in the second movement of Beethoven, you can very clearly hear the three-note call of the European quail and the two-note call of the cuckoo.

And there's Mozart's starling. Meredith West and Andrew King of the Indiana University have studied Mozart's piece *The Musical Joke,* which musicologists have been puzzled by, because it's not really of the quality of Mozart. The horns are playing off key and there are all sorts of strange devices there that offend musicians. West and King kept starlings, and they found that starlings do all the things that are found in this piece by Mozart, such as stopping a motif in midsection. And this particular piece was written six days after his pet starling died. Mozart wrote it, we can surmise, as a tribute to his friend.

___ *I've heard birds sing in what seems like a sonata style. Can they?*

They do, in the sense that they have an exposition of a theme. Very often, they have variations in theme reminiscent of canonical variations like Mozart's Sonata in A major, where you have theme and variation. And eventually, they come back to the original theme. They probably do it for the same reasons that humans compose sonatas. Both humans and birds get bored with monotony. And to counter monotony, you always have to do something new to keep the brain aroused.

— *Bird song patterns are quirky and cute, but what do they mean?*

Well, it doesn't have to mean anything. It's just wonderful that animals and humans can converge.
May 16, 2000

Postscript

I discovered Luis Baptista at the annual meeting of the American Association for the Advancement of Science in the winter of 2000. He was giving a paper on bird dialects and song. I was enchanted. There was something so alive, so original, so sweetly engaged about Baptista I knew he'd make a fabulous interview subject.

A few weeks after the conference, I jetted out to San Francisco to spend a day with him. We spent some time in Golden Gate Park where he seemed to know the particular life story of every single creature, had lunch at Cliff House where we said "hi" to the seals, and then we went off to another park in Haight-Ashbury where there were a group of escapee green parrots living in the trees.

So often when I do an interview, I learn something new—not just about the subject matter in question, but how to live, what passion means, or the value of work. And from our day together, I the city-born woman, learned something I'd always suspected: that life is incomplete without a closeness to nature.

A few weeks after the interview appeared, I found a message on my voice mail from an Amy Kramer, who was in public relations at the California Academy of Sciences, where Dr. Baptista worked. There was something in her tone that made me dread phoning back. "Luis is dead," she said when we finally did connect. "They don't know what happened—probably a heart attack."

And so I found myself producing my first obituary for *The New York Times.* "Luis F. Baptista, a leading expert on bird song who had learned to recognize not just bird languages but also dialects and regional accents, died on Monday at his home in Sebastopol, California," I wrote. "He was fifty-eight.

"Dr. Baptista was tending to a wild barn owl that had moved into a shed on his property when he collapsed, said his companion Helen Mary Horblit. The cause of death is not immediately known. . . ."

In March of 2001, nine months later, I phoned Helen to see how she was doing. She'd sold the ranch where they'd lived together and was devoting herself to finishing a project Luis had started.

In the closing months of his life, he had developed a plan to reintroduce the soccoro dove—extinct in the wild, but existing in captivity—to its native habitat in the Revillagigedo Islands, 400 miles off the west coast of Mexico. "It's going to happen," Ms. Horblit reported. "We have the money to build the aviaries and the new Mexican government is very supportive. The symbolism of the project makes me want to be alive—to save a part of the world is a privilege. I would have really lost it if I didn't have this to do in Luis' name."

DARWIN'S PROGENY

STEPHEN JAY GOULD

Primordial Beasts, Creationists, and the Mighty Yankees

Rick Friedman

It was a sunny afternoon in SoHo and the paleontologist **Stephen Jay Gould**—president of the American Association for the Advancement of Science, Vincent Astor visiting research professor of biology at New York University, and the Alexander Agassiz professor of geology at Harvard—was sitting in his loft, ruminating about the pleasures of finally living in Manhattan.

Dr. Gould, 58, has spent much of his life circling Manhattan. He grew up in 1950s Queens in a working-class family at a time when Manhattan was the ever-distant "city." In 1967, Dr. Gould got his Harvard appointment, which meant, of course, living in Cambridge and being one of the few Yankees fans in all of the Harvard Yard.

Four years ago, Dr. Gould, who was divorced, married a sculptor and art historian, Rhonda Roland Shearer of Manhattan, now 45, and together they set up housekeeping in SoHo, in a vast urban spread filled with Tiffany lamps, good art, and first-edition scientific tomes.

In his 19 books and in essays for *Natural History* magazine, Dr. Gould has become perhaps the most eloquent and best-known proponent of the view that evolution and natural selection are responsible for the origin and diversity of species. But earlier this month he came under criticism in *The New Yorker,* which suggested that his emphasis on chance in the evolutionary process had unwittingly aided the cause of creationism. Dr. Gould declined to respond to *The New Yorker* article, by journalist Robert Wright, saying that he did not believe that such personal attacks merited a response and that his work spoke for itself.

The Harvard paleontologist did, however, speak about other aspects of the ongoing political struggle between creationists and evolutionists.

— *What was your reaction, when you first read that the Kansas Board of Education was going to make the teaching of evolution optional in biology classes?*

That the citizens of Kansas would be profoundly embarrassed by the stupidity of the ruling and that they would vote that school board out of office the next year. The Kansas school board's decision is absurd on the face of it. It's like saying, "We're going to continue to teach English, but you don't have to teach grammar anymore."

But the creationists can't do what they want to do because of the history of Supreme Court decisions. They are very restricted in terms of a legally defendable stand. This is probably the only thing they can do.

The only reason it happened is that nobody votes in school board elections anymore. Thus, determined minorities can take over. It took this fundamentalist group three election cycles to take over in Kansas. They only have a one-vote majority. Four are up for election next year.

The bigger dangers aren't these legal maneuvers. It's the thousands of teachers who are less than optimally courageous, as most humans are, who are probably teaching less evolution because they don't want trouble. You can't even measure that.

___ *Is creationism a uniquely American phenomenon?*

That's not hard to see. It just doesn't happen any place else in the western world. Europeans just don't get why we have it. There are two things that European intellectuals don't understand about Americans, I find. One was Bill and Monica, or, our obsession with it. The second is how you can possibly have an anti-evolution movement in a modern scientific country.

___ *There is a recent trend in the social sciences to go to neo-Darwinist explanations of social problems: a kind of mutant resurgence of the Social Darwinism of the late nineteenth century. Why has this happened now?*

This is a conservative age and I think, it's tempting for conservatives to argue, "Why are you calling for change or equalization when what we have now reflects the natural state of human nature?"

Also, I think, we sometimes make a misuse today of Darwin in terms of trying to assuage our disappointments with some of our worst traits. That is, if we don't like our aggressivity or our sexism, we might try to fob it off with: "Oh, well, we're made that way. We can't help it."

___ What about the appeal of neo-Darwinism to people who like their traits? The biological explanation "it's a gene" has, for instance, become very popular with gay rights advocates.

Oh yeah. This is an age that largely, wrongly, I think, favors genetic explanations. So it's going to spread everywhere. But I think that's a two-headed argument. Because if you put your eggs in that basket, then suppose it turns out that you're wrong? You don't want to base a defense for a defendable bit of our diversity upon its putative biological nature.

I'd rather take the point of view that it has nothing to do with the biology. It's an ethical issue.

___ As someone who publishes in both scientific and popular media, what's your take on the quality of academic writing?

Compared to what? I don't think academic writing ever was wonderful. However, science used to be much less specialized. There wasn't much technical terminology, and then, most academics are not trained in writing. And there is what is probably worse than ever before, the growing use of professional jargon.

And I think it arises more out of fear than arrogance. Most young scholars slip into this jargon because they are afraid that, if they don't, their mentors or the people who promote them won't think they are serious. I can't believe that anyone would WANT to write that way.

___ Do you think your colleagues sometimes resent you because you have, horror of all horrors, penned a few bestsellers?

Oh, sure. Anyone who has success in writing for the general public is envied. Goethe died in 1832. As you know, Goethe was very active in science. In fact, he did some very good scientific work in plant morphology and mineralogy. But he was quite bitter at the way in which many scientists refused to grant him a hearing because he was a poet and therefore, they felt, he couldn't be serious. This is not entirely a new phenomenon.

___ Do you write easily?

I don't know what writer's block is.

— *What does writing do for you?*

It's the best way to organize thoughts and to try and put things in as perfect and as elegant a way you can. A lot of scientists hate writing. Most scientists love being in the lab and doing the work and when the work is done, they are finished. Writing is a chore. It's something they have to do to get the work out. They do it with resentment. But conceptually to them, it is not part of the creative process. I don't look at it that way at all. When I get the results, I can't wait to write them up. That's the synthesis. It's the exploration of the consequences and the meaning.

— *Since your marriage to Rhonda Roland Shearer, you have been living half-time in New York and half in Cambridge. To what extent has this new life left you feeling split?*

The big frustration is waiting for this decent train service between Boston and New York to start. But I like living in New York, though I don't feel that I ever left. I grew up in Fresh Meadows, went to Jamaica High School.

— *You didn't go to Bronx High School of Science?*

It was too far. I got on a bus and subway and it took me two hours to get there, and I thought, "I'm not going to spend four hours a day for the next three years on the subway." So I went to Jamaica High School. You know, New York had a great public school system once and it will again, I trust. I feel I got a great education at Jamaica High. And P.S. 26 before that. I'm nothing but an old city kid at heart.

— *Your recent book,* Questioning the Millennium, *was, among other things, a lengthy investigation of Year 2000 issues. Tell us, are you and Rhonda secreting bottles of water and cords of firewood for fear of what will happen when the clocks change?*

No, there's been a lot of attention to Y2K and a lot of testing. I don't expect any. As a matter of fact, I will be singing in a concert of Haydn's *Creation* in Boston on New Year's Day. I'm going to have to get from here to there for rehearsal. I will drive up there, though.

I don't think anything significant is going to happen. Insofar that there are some worries on a global scale, the things I would worry about are places that are really cold, like northern Russia, where there could be an interruption of the electricity and heating and things like that.

The funniest thing you can say about it all is that in the year 1000, insofar as people were aware of the millennium, their fears were grander. They feared the apocalyptic revelations of Revelations. They really thought that Jesus would come again, that Satan would be bound and the world, as we know it, would end. I think it's so amusing that in a secular age the main fear that people have is caused by a technical glitch caused by a computer misreading a date because of poor anticipation by some programmers 30 years ago.

December 21, 1999

Postscript

After heading the American Society for the Advancement of Science, Stephen Jay Gould has spent the past year finishing up his magnum opus for Harvard University Press, which he describes as "a twenty-five hundred page work on the structure of evolutionary theory." The study will likely be published in 2002 in tandem with his tenth book of essays, tentatively titled, *I Have Landed,* which are the words his grandfather wrote into the first book he ever purchased after coming to America. "What I've been up to," Steve Gould told me via telephone from his office at Harvard, "is much of the same—teaching, writing, getting back to more technical research, getting back to the field. Not that I ever stopped doing it. To study the paleontological history of the last two thousand years—the cerion, the snail."

When last we'd spoken, Dr. Gould was chagrined about the difficulties of commuting between his two posts in Cambridge and New York. Since that time, the much-awaited Acela bullet train has come on line. "It's a joke by any European standard to call it a fast train, but it is a beautiful train, it's a lovely ride," the palentologist commented about this new technologic improvement in his life. "It makes all the difference and I get three and a half hours of good solid work. It's much better than flying, which is forty-five minutes and all kinds of things can go wrong. I'm a worry wort. I hate missing lectures!!! The one downside to the Acela is those cell phone people. THERE MUST BE A CELLPHONE FREE CAR!!!! It will happen."

FRANCISCO J. AYALA

Ex-Priest Takes the Blasphemy
Out of Evolution

Darcy Padilla

Francisco J. Ayala, former Dominican priest, present day wine-grape grower, art collector, author of 12 books and 650 articles on genetics, and a professor of biological sciences and philosophy at the University of California at Irvine, is known in the science world as the Renaissance man of evolutionary biology.

Dr. Ayala's many interests are in evidence at his office in Irvine on a recent morning. Paintings, sculptures, and good carpets decorate the room—as well as a collection of stuffed parrots, which Dr. Ayala picked up some years ago while on an expedition in the Amazon.

—— *You teach a basic biology class here at the University of California at Irvine. How do you handle students with religious objections to the theory of evolution?*

I treat them with respect, and I try to get a dialogue going. Every year, I start out my course with the theory of evolution. So on Day 1, there's always a line of students complaining, "Professor Ayala, I am only here because I want to go to medical school—I cannot accept evolution because I am Catholic."

With Catholics, I take out the Pope's address to the Pontifical Academy of Sciences in October 1996 where he endorses evolutionary teachings. If the students are Christian fundamentalists, I tell them that there are many Protestant theologians who agree with evolution. I say that evolution, in my view, is not only not anti-Christian, but the idea of special design, which many fundamentalists adhere to, might be—because it teaches the view of God that is blasphemous. The Special–Design–God is a God who messes up. Think about all the backaches, impacted wisdom teeth and painful childbirth that exist because we humans evolved incompletely! "Do you think God is absent-minded?" I ask them.

—— *Why do you think creationism has such a strong hold on America?*

It is rooted in historical origins. This is a country colonized originally by people seeking their own religious feelings, people who felt themselves marginalized because of them. This beginning gave rise to a kind of religious populism, which remains a strong part of U.S. culture. Even today.

___ *There are people saying that the evolutionary process has come to an end because of advances in genetic engineering. Do you agree?*

Not for a moment. Genetic engineering can accomplish many things in agriculture, but it will probably not have as great an effect on the near future on humanity. And evolution will continue. Indeed, one can demonstrate scientifically that biological evolution continues, even now, in modern humans. In fact, it is accelerating because of the rapid changes in the environment.

Technology is changing the world we live in, faster and faster. That is prompting our natural evolution to occur faster and faster. Yet, it is still very slow. The evolution that is really important in humans comes through cultural change. In modern times, we do not evolve so much by slowly changing our genes as we do by rapidly adjusting the environments to the needs of our genes. For example, our genes make us adapted to live in tropical climates and yet, we have colonized the whole planet. And we've done it, not by changing our genes, but by using clothing and shelter.

___ *You're not worried about cloning. Why?*

I think the general public has a misunderstanding of what cloning means. Contrary to what people think, a human being cannot be cloned. What can be cloned are genes.

Yes, one can clone the genes of Francisco Ayala. But the human being that will develop will be a completely different person from me. He may look somewhat like me at the same age, but he will not have gone through a sequence of experiences that started in my mother's womb, continued in my family in Franco's Spain, and that went on with me during my emigration to America. The individual that is developed from these same genes will experience a completely different environment and will be different in everything that counts—temperament, personality, experience.

— There's a colleague of yours here at Irvine, Dr. Michael Rose, who's doing genetic experiments aimed at increasing the longevity of fruit flies. You've been somewhat unhappy with him. Why?

I'm not unhappy with him personally. He has been very successfully studying the genetics of longevity and he has selected strains of flies that live longer than the average. Alas, he's accomplished that by killing off the flies that are not so long-lived and doing that over many generations. My problem is with the way Rose goes about it: he's much too drastic—he lets 99 percent of the flies die. I don't think we'd want to use his methods to increase human longevity.

— Sounds like you have a great fondness for drosophila, *fruit flies?*

I do. They are not ugly or anything like that, which people think. Quite to the contrary! When you look at them under the microscope, they are very, very attractive creatures. Also, they are very useful for studying genetics. In a culture you can grow several hundred of them at very little cost and in just a few days. On the other hand, much as I like them, I have no problem with grinding them up for their DNA.

— You have worked as a member of the National Advisory Council for the Human Genome Project. Do you think that when the project makes its report, we will know everything we need to know about human genetics?

Oh, no. Far from it. We will know some things that will be helpful. However, we will still be a very long way from understanding what we humans are, or even how our genome works. One can know the sequence of letters in the *Encyclopedia Britannica,* but that doesn't give us the knowledge contained inside of it.

— What exactly was your role within the Genome Project?

One of my concerns was that human diversity be taken into account. We should not assume that when we have a human genome mapped that it is the genome of the entire race; every human being is different. This is something that one has to keep alerting people about again and again: that we are getting the map of one individual, not the map of the human race.

I pushed for the dedication of some of the budget to evaluate the ethical and legal and societal implications of what would come once we had the map. It was early on decided that 3 to 5 percent of the budget of the whole project would be dedicated to ELSI, which is an acronym for Ethical, Legal, and Social Implications of the Human Genome Project.

___ *You were a part of a scientific group that discovered that all the world's malaria parasites were descended from one original creature. What are the implications of that finding?*

That malaria spread throughout the world comparatively recently—after the onset of agriculture. It also implies, very importantly, that since all the malaria parasites are virtually identical in their DNA, that there is the possibility of finding a magic bullet for the disease.

___ *When a new creature is discovered somewhere in the world, what do you feel?*

If it is an important new species, I get very very excited. That's what happened when a class of organisms was discovered about 15 years ago—Archaea. These are just like bacteria, but they are completely different from them and from all other microorganisms. And they have opened up a whole series of questions to investigate. Learning about them was comparable to the first time I saw *Guernica* by Picasso when I came to New York in 1961. It impacted me in every important way. It was a sense of opening up new ways of looking at the world.

___ *Congressman George E. Brown Jr. recently told me he found scientists to be really poor at getting their needs onto the national agenda. As the former president of the American Association for the Advancement of Science, do you have insight into why this is?*

My experience is consistent with his. I think it's because we're parochial. That's the essence of being a scientist. You have to focus all your energy into some problem to solve it. In the process, you become so narrow-minded that you lose perspective of what else is out there. Scientists are, I think, zealots. They are completely focused on what interests them, and they pay little attention to anything else.

___ *What ever prompted you to buy a vineyard in northern California?*

First, it's an incredibly beautiful place, 400 acres at a high point on the Mokelumne River. I can write there, spend weekends there. And second, I thought it could be a wonderful business, which it has been.

— *Do you use any of your scientific training in the grape business?*

I use my scientific training to make decisions. For instance, when I bought the vineyard in 1981, I was told by very experienced wine producers that chardonnay grapes would not grow well in that region. I was not persuaded. I planted many acres with the usual grapes and four with chardonnay. And the chardonnay did very well. So I planted 163 more. Now, I use the 400 acres to make some wine for myself, and the rest I sell to Mondavi, Sebastiani, and one or two others.

— *Have you considered selling your wine under a special "Darwin" label?*

No, but it's a very good idea.
April 27, 1999

Postscript

When I interviewed him in his office at the University of California in Irvine in February of 1999, I thought that Dr. Ayala had the grandest and most beautiful workspace I had ever seen in an academic environment. There were good rugs on the floor, interesting art on the walls, and everywhere were these fantastic stuffed birds that Ayala had picked up on some long ago safari into the tropics. The room looked as if Frida Kahlo had been his decorator.

Indeed, the office screamed its owner's personality. (It said: I am of the world and enjoying all the beauty in it.) I liked that about Ayala. Unlike so many one meets in the academy, his universe was huge. He could talk about genes and God and gods and the great God of wine, with the enthusiasm of a man fully engaged in the beauties of the world. (I once attended a lecture he was giving on evolution, which he opened with a slide of Picasso's epic painting of the Spanish Civil War, *Guernica*.)

Thus, I was surprised to hear when I phoned Dr. Ayala in March of 2001, some two years after our interview, that there was "nothing really new" in his life. I pressed him, asking the stock question that I'd put to almost all the subjects in this book: Had he done any interesting work for lay people in his area recently? Across the continent from my homebase in New York, I could hear

Dr. Ayala's voice brighten. Yes, he was producing a weekly column on scientific issues for the Spanish newspaper, *El Correo*. And yes, he was also producing a series of popular books to be sold both in Iberia and Latin America on scientific topics. But on the whole, "nothing new."

Our conversation was short. Formal. A bit disappointing. "Write me more," I suggested. And a few days later, came an e-mail, which was much more in character:

"Dear Claudia:

"It was nice to hear again from you.

"Something that may be of interest to you is that I do numerous writings in Spanish addressed to the general public. There are good science journalists in the U.S. and U.K. and good popular books on evolution, genetics, etc., but many fewer in Spanish. Thus, I write from time to time articles for the two most widely distributed dailies in Spain, *El Pais* (political left) and *ABC* (right). Currently, I am writing a series of articles on the human genome, which appear successively in the Sunday edition of *El Correo* (published in Bilbao, and the largest circulation daily in the Basque region. Remember: My genes are 7/8 Basque).

"I also have written in Spanish over the years a number of general-audience books with titles like: *Unfinished Nature, The Theory of Evolution, Origin and Evolution of Humankind*—the Spanish market is huge, because it includes Latin America (largely dominated by the great Spanish publishers). So these books appear in several disguises and reprints and are all still in print (one for more than 20 years!).

"These writings, as well as the public lectures I give in Spain, have a positive impact on my art collection. All royalties and honoraria get readily invested in purchasing works by contemporary Spanish painters. Last November I received a windfall and gave myself a wonderful Christmas present of five beautiful oil paintings and one watercolor. Hana greatly enjoys the art, but I suspect she enjoys even more the pleasure I get from it, including purchasing it."

BIRUTE GALDIKAS

Scientist at Work; Saving the Orangutan, Preserving Paradise

Aurelija Mituziene

Of the three young women recruited in the 1960s by the paleontologist Louis Leakey to study great apes, **Birute Galdikas,** now 53, is the least known. Dr. Leakey's first disciple, Dr. Jane Goodall, who discovered that chimpanzees made tools, has become an international scientific celebrity, and Dr. Dian Fossey, who lived among the gorillas of Rwanda and was killed there in 1985, was played by Sigourney Weaver in the movie *Gorillas in the Mist.*

But the story of Dr. Galdikas, who quietly devoted herself to the study and preservation of the Indonesian orangutan, remains largely unknown.

"That's because I have a name nobody can pronounce and because I've been in Borneo all these years, tracking an elusive and solitary animal," Dr. Galdikas, whose name is Lithuanian (pronounced bi-ROO-tay GALD-i-kus), said on a recent morning. She had come to New York for several days to take the middle of her three children, Jane Galdikas, 15, to museums and the theater.

Still a resident of Borneo, Dr. Galdikas recently became an Indonesian citizen because "the orangutans are Indonesian and because someone in the government suggested it would be helpful."

But she flies regularly to North America to teach at Simon Fraser University in Canada and to oversee her charity, the Orangutan Foundation International of Los Angeles.

— *Give us a report on the state of the world's orangutans.*

They are poised on the edge of extinction. It's that simple. We're still seeing orangutans in the forest; they are coming into captivity in enormous numbers. You just know that there can't be that many left in the wild.

— *How did the orangutans come to be so threatened?*

77

The main factor was that until 1988, Indonesia had a forestry minister who was a real forester. In 1988, he was replaced by a forestry minister who was an agriculturist, a promoter of plantations. That signaled a shift in government policy from selective logging to clear-cutting of the forest. For orangutans, clear-cutting is a policy of extinction. If you selectively log, some animals will survive. But with clear-cutting, the habitat is gone. If that weren't enough, in 1997, there were these horrendous fires that devastated the forests.

Moreover, the last three years have been a period of intense political upheaval: an economic crisis, ethnic strife, student riots, President Suharto's resignation. After President Suharto stepped down in 1998, there was a vacuum of power in the center. Once people in the provinces understood that, some felt they could do whatever they wanted. And what some of them wanted to do was log the forest. So throughout Indonesia, places that had, more or less, been protected, became besieged.

At first, only local loggers came in. When nobody stopped them, the bigger commercial loggers followed. Suddenly, there were no more protected parks.

— *Is this true too in Kalimantan, Borneo, where you have your research station?*

Yes, though in the National Park where I work, we're doing what we can. We're trying to set up patrols of local men to go out with park rangers so that when they come across illegal loggers, they don't feel totally intimidated. We're working with the Indonesian government to set up new wildlife reserves at expired logging concessions. And of course, we're doing what we always have: saving wild-born orangutans who've been captured by humans.

We have a hospital for 130 orangutans. We have an orphanage for the babies. Eventually, they are released to the wild, though with the fast-disappearing habitat, it's always tough to find a safe place for them.

— *Tell us what you've learned about orangutans in the nearly 30 years you've been studying them.*

Well, we've gotten a picture of a very long-lived primate who probably lives 60 to 70 years in the wild. They use a wide variety of foods in the wild, about 400 different kinds, because food is generally scarce for them.

The males come and go. They're very, very competitive. Probably very few males are successful at actually impregnating females. And the females seem to get pregnant about once every eight years. Also, they're very smart. When orangutans have interactions with humans, they use tools at an incredibly rapid pace.

— *Based on what you've seen, do you believe that orangutans can learn language?*

I think orangutans can learn how to use language at the level of a 3-year-old child. I had a student in 1978, Gary Shapiro, who came to Camp Leakey, and he taught an adult female, Rinnie, sign language. He could not believe how fast she learned it. Rinnie took the tutoring personally. One day, Rinnie took Gary by the hand and tried to seduce him. Gary pushed her away. She thereafter lost all interest in signing. Interestingly, my son, Binti, who was then 2, picked up signing from watching Gary and Rinnie together—though Binti thought that you could communicate with all orangutans through signs. For a while, he went around and signed with all the animals, even those who'd never been taught sign language.

— *Did Binti speak orangutan?*

He could interpret what they meant. He moved it and he felt it. His whole body posture would be like an orangutan.

— *Did Binti identify as an orangutan?*

It was heading in that direction. A friend, a psychiatrist, came to Camp Leakey in 1979 when Binti was 3 and said, "You know, you should get him out of here. He really needs to be with kids his own age and to go to preschool." I knew that Jane Goodall's son, Grub, had been through something similar, and they sent him to boarding school. So I sent Binti to live with my ex-husband, his father, Rod Brindamour, in Canada. Letting Binti go with his dad was very, very hard. But he had no people his own age and size at Camp Leakey. All he had were orangutans and gibbons. Binti's first friend was a gibbon and she really liked him. This was when he was a baby in a crib.

In fact, Binti right now is in Indonesia, working as a volunteer. He's 23 and doesn't want to do this work professionally, but he does want to help.

Neither of my other kids—Jane, 15, and Fred, 17—wants to do this, either. They see how hard it is.

_ *How did you meet your second husband and the father of your two younger children, a Dyak named Pak Bohap?*

He worked at Camp Leakey. The first time I laid eyes on him, I saw him in profile, and it was like I was punched in the stomach. I couldn't take my eyes off of him. He was such a handsome man. I was driving and I almost went into a ditch. I was married at the time. After that, I avoided him. I just didn't want to go close.

_ *How were you able to transcend the barriers of race and culture to make a marriage with a Dyak?*

It was possible because the Dyak are probably the most egalitarian people on the planet in the ways of men and women. This is what an anthropologist told me long before I married Pak Bohap. The astonishing thing to me is how comfortable we are. And the fact that we come from different religions and world views actually keeps the marriage fresh and interesting.

_ *Are you still together?*

Oh yes. He hates it that I travel so much.

_ *What's your take on this revival of Darwinism in the social sciences?*

I don't like the misuses of it. The interaction between environment and genes is so complicated that to use genetic arguments, one has to be careful. Like the recent thing about rape now—where people are saying that human rape is not about violence, it's about sex, it's just a reproductive strategy—that's baloney. Maybe it is about reproductive tactics, but it's also about social control.

_ *Orangutans rape, don't they?*

In Orangutans, rape is about sex. That's all it is. I mean, they don't beat up the female. They don't kill her. Once it's over with, it's over with. Most orangutans that are raped are of childbearing age. With humans you have

90-year-old nuns and 3-year-old girls getting raped. So, there's something other than sex going on with humans.

— *Getting back to orangutans, when you see what is happening to the animals you've spent your life studying, what do you feel?*

I feel like I'm viewing an animal holocaust and holocaust is not a word I use lightly. The machine of extinction is grinding away. The destruction of the tropical rain forest in Borneo is accelerating daily. The consequences of this destruction for the orangutans will be final. And if orangutans go extinct in the wild, paradise is gone. And we'll never have it again.
March 2, 2000

Postscript

In the year since I've seen her, Birute Galdikas' existence has been filled with death—the death of her father, her brother-in-law from pancreatic cancer, the ongoing death of the great apes of Indonesia. Between all that, she is living in three different worlds—in Canada where she supports her family by teaching at Simon Fraser University, in Los Angeles where her children and her foundation are based, and in Borneo with the animals. Hers is not an easy life.

Yet, when I speak to her on the telephone in March of 2001, her tone is cheerful, optimistic. She says that she manages to still spend six months every year in Indonesia and that, "in the northern part of the park where I work, through the Orangutan Foundation International we have organized blockades of the river mouths and we have almost stopped illegal logging and goldmining—except in an area where there are community land claims."

Indonesia, Dr. Galdikas reports is, "very unstable and chaotic. It's still very much a former dictatorship transitioning to democracy and that is affecting the animals terribly because there isn't the political will to enforce the laws of the nation. People are doing what they want, and what they want is to exploit the natural resources."

The one highlight of last year, Galdikas claims, was the passage in the United States of the Great Ape Conservation Act, which provides money for conservation from the U.S. government. Dr. Galdikas managed to lobby for the law from Indonesia by placing calls to various U.S. Senators, and she jokingly says, "that made more of an impression than if I'd flown into Washington and seen them in person."

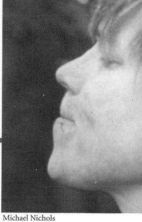

EMILY SUE SAVAGE-RUMBAUGH

She Talks to Apes and, According to Her, They Talk Back

Emily Sue Savage-Rumbaugh, 52, a researcher at Georgia State University in Decatur, Georgia, studies communication among primates and runs a 55-acre laboratory near Atlanta where she trains animals and humans to communicate with each other.

Dr. Savage-Rumbaugh is the author of *Kanzi: The Ape at the Brink of the Human Mind,* and with Stuart G. Shanker and Talbot J. Taylor, is a co-author of *Ape Language and the Human Mind,* to be published next month by Oxford University Press.

— *Do your apes speak?*

They don't speak. They point to printed symbols on a keyboard. Their vocal tract isn't like ours, and they don't make human noises. However, they do make all kinds of ape noises. And I believe they use them to communicate with one another. Now, the apes may not always elect to talk about the same things we do. They might not have a translation for every word in our vocabulary to theirs. But from what I've seen, I believe they are communicating very complex things.

Let me give you an example. A few weeks ago, one of our researchers, Mary Chiepelo, was out in the yard with Panbanisha. Mary thought she heard a squirrel and so she took the keyboard and said, "There's a squirrel." And Panbanisha said "DOG." Not very much later, three dogs appeared and headed in the direction of the building where Kanzi was.

Mary asked Panbanisha, "Does Kanzi see the dogs?" And Panbanisha looked at Mary and said, "A-frame." A-frame is a specific sector of the forest here that has an A-frame hut on it. Mary later went up to "A-frame" and found the fresh footprints of dogs everywhere at the site. Panbanisha knew where they were without seeing them.

And that seems to be the kind of information that apes transmit to each other: "There's a dangerous animal around. It's a dog and it's coming toward you."

— *Your apes watch a great deal of TV—why?*

Because their lives are so confined. They can expand their world by watching television.

— *What do they watch?*

This varies. They like the home videos we make about events happening to people they know from around the lab. They like suspenseful stories, with an interesting resolution. Of movies we buy, they really like films about human beings trying to relate to some kind of ape-like creatures. So they like *Tarzan, Iceman, Quest for Fire,* the Clint Eastwood movies with the orangutan.

— *You have a game with the apes, "Monster," where a lab staffer dresses up in a gorilla suit and feigns being frightful. Why?*

It's a game started some years ago when we were working with two chimps, Sherman and Austin. We discovered that if someone dressed up in a gorilla suit and we drove this "monster" off with poundings of hammers and sticks, we upped our status with the chimps. In other words, "We're not the experimenters in charge. We're your helpers." Sherman and Austin didn't know we were playing. For a while Kanzi and Panbanisha didn't either. But they caught on soon enough and now they love the game. . . .

Another time, Panbanisha and I were walking around the building where Sherman and Mercury, this male chimpanzee with a big interest in Panbanisha, lives. Mercury came outside and was being really bad—displaying, throwing bark, and spitting at Panbanisha. So Panbanisha opened her backpack, where there was a gorilla mask inside and she pointed to symbols on the keyboard and asked Mary to play Monster. Mary did that, and Mercury flew indoors.

Panbanisha was able to use the game to stop him from displaying at her. She knew it was pretend. He didn't.

— *How do you know when the chimps point to symbols on the keyboard that they are not just pointing to any old thing?*

We test Kanzi and Panbanisha by either saying English words or showing them pictures. We know that they can find the symbol that corresponds to the word or the picture. If we give similar tests to their siblings who haven't learned language, they fail.

Many times, we can verify through actions. For instance, if Kanzi says "Apple chase," which means he wants to play a game of keep away with an apple, we say, "Yes, let's do." And then, he picks up an apple and runs away and smiles at us.

— *Some of your critics say that all your apes do is mimic you.*

If they were mimicking me, they would repeat just what I'm saying, and they don't. They answer my questions. We also have data that shows that only about 2 percent of their utterances are immediate imitations of ours.

— *Nonetheless, many in the scientific community accuse you of over-interpreting what your apes do.*

There are some who say that. But none of them have been willing to come spend some time here. I've tried to invite critics down here. None have taken me up on it. I've invited Tom Sebeok (of Indiana University) personally and he never responded. I think his attitude was something to the effect that, "It's so clear that what is happening is either cued, or in some way over-interpreted, that a visit is not necessary." I would assume that many of the people associated with the Chomskyian perspective including Noam Chomsky himself have the same approach: that there's no point in observing something they're certain doesn't exist.

Their belief is that there is a thing called human language and that unless Kanzi does everything a human can, he doesn't have it. They refuse to consider what Kanzi does, which is comprehend, as language. And it's not even a matter of disagreeing over what Kanzi does. It's a matter of dis-agreeing over what to call these facts. They are asking Kanzi to do everything that humans do, which is specious. He'll never do that. It still doesn't negate what he can do.

— *Your husband, Dr. Duane M. Rumbaugh, is a distinguished comparative psychologist who is a pioneer in the study of ape language. Has your research been helped by the fact that your personal life is so fused with your profes-sional life?*

Without our being together, I don't think that one could ever be responsible for as many apes as we have here. Duane and I live right near the research center, and we're willing to go there day and night, 365 days a year. If an ape is sick, if one of the apes has gotten free, if Panbanisha is frightened because she's heard the river's about to flood, we go.

There have been lots of frictions, though. Duane was very, very upset when I began taking the apes out of their cages. And when I began to say that Lana (Duane's chimp) didn't understand some of the things she was saying and that comprehension of language was important, not just production—we almost broke up over that.

But we really love each other, and we're united in our core beliefs: that there is a huge capacity on the part of apes and probably all kinds of other animals that's being ignored. By ignoring it, humans are separating ourselves from the natural world we've evolved from. The bonobos are a real bridge to that world. At base, no matter how much Duane and I argue, we both know this is true.

April 14, 1998

Postscript

Since I conducted the interview with Dr. Savage-Rumbaugh, there have been huge changes in the lives of the Georgia State bonobos—and also in the life of Emily Sue Savage-Rumbaugh.

On a personal level, Savage-Rumbaugh and her husband, Duane Rumbaugh, separated briefly. They are now back together again. For the apes, there was a baby boom. Panbanisha had a boy, Nyota, and recently, another, Nathan. Matata, who was born in the wild and who does not have communicative skills with the symbolic keyboard, had a girl, Elykia, and later a boy, Maesha.

This population explosion was due, in part, to the importation to the Language Center of a male bonobo, who'd been raised with chimpanzees as part of a Japanese circus troupe. Once P-Suke got over acting like an aggressive dumb chimp, the female bonobos really took to him and he's become a major stud.

After Nyota was born in April of 1997, Savage-Rumbaugh moved into the area of her laboratory where Panbanisha lived; she wanted to see if the infant might develop greater language skills if he was raised by his ape mother and also by a human mother. The experiment continued apace until one day Kanzi, one of the male bonobos—the first one to have language—bit one of

Nyota's keepers. Till that time, there had been two groups of apes living separately at the Georgia State facility: those with language and those without.

The attack forced Savage-Rumbaugh to combine the ape groupings and stop living with Pambanisha and her offspring. Her research project has, thus, changed.

"Why did Kanzi bite someone if he understood language and was so attached to people? He did so for the same reasons that some human beings attack friends and relatives when they feel they have been disgraced or compromised," Savage-Rumbaugh wrote me in February of 2001. "Kanzi felt that his human friend should have taken actions when a human visitor to the lab behaved in a manner than Kanzi deemed was inappropriate and aggressive toward me. When this did not stop, Kanzi told another researcher to bite the visitor. When this did not occur (as of course it could not) Kanzi took the next opportunity to immediately bite the person he felt had engaged in traitorship.

"The more time we spend with apes the more we are expected to adopt their rules, their culture—and the more transgressions are punished. It was not that we were in any danger being with them, or that they were trying to become dominant. It was rather that we were assigned social roles and it was our duty to carry them out—in and out of the cage, regardless of what human needs or human interpretations dictated as appropriate actions.

"In the future we will be focusing on interpreting the vocalizations of the bonobos and attempting to determine if they can translate some of these for us. We will also look at comparisons between Elykia and Nyota on a battery of tests since Elykia was reared by her wild-born mother. And Nyota was reared by myself, Bill, Dede, and Panbanisha. We will be looking at the neuropsychological differences between the two infants. We hope that many other researchers will become interested in investigations with these apes as well, due to their unique rearing backgrounds."

ANNE FAUSTO-STERLING

Exploring What Makes
Us Male or Female

Rick Friedman

Providence, Rhode Island

On a recent frozen winter evening, **Anne Fausto-Sterling,** 56, a professor of biology and women's studies at Brown, sat in a restaurant here, nibbling on a light snack and talking about her favorite subject: the application of ideas about gender roles to the formal study of biology.

In the academic world, Dr. Fausto-Sterling is known as a developmental biologist who offers interesting counterpoints to the view that the role division between men and women is largely predetermined by evolution.

"When people say 'it's nurture' or 'it's nature' in making us male or female, I take the middle ground and say that it's a combination of both," she said. "That's not a popular position to take in today's academic environment, but it is the one that makes the most sense."

Her 1985 book, *Myths of Gender: Biological Theories About Women and Men,* is used in women's studies courses throughout the country. Dr. Fausto-Sterling's newest work, *Sexing the Body: Gender Politics and the Construction of Sexuality,* is a look at societal ideas about gender as seen through the eyes of human beings defined as neither male or female—hermaphrodites.

Until 1980, she studied the role of genes in the embryological development of fruit flies. More recently, she has investigated the developmental ecology of flatworms.

__ *What can we learn about gender from examining how the medical profession treats infants born with ambiguous genitalia? These are children who were once called "hermaphrodites," and whom you would prefer we term "intersexuals."*

From them, we can literally see how society's ideas about male and female are reconstructed. When infants with ambiguous genitalia are born, everyone—parents, doctors—is very upset and the physicians often suggest drastic surgeries to assign a specific gender to the child. The regimen usually involves the doctors' deciding what sex the child ought to be.

Then, they surgically reconstruct the patient to conform to that diagnosis: body parts are taken out, others are added, hormones are given, or taken away.

In the end, the doctors take a body that was clearly neither male or female and turn it into one they can represent to the world as "male" or "female."

— *How did the fate of intersexual children become your passion?*

In the early 1990s, I began looking into this because I was interested in a theoretical question that was circulating around feminist studies at that time; I wanted to know, What is meant when we say, "the body is a social construction"? At the time, social scientists were looking into how our ideas about the human body were shaped by politics and culture. That inquiry led me to a lot of the medical literature on intersexuality.

— *How many people do you estimate are born intersexuals?*

It depends on how you count. Working with Brown undergraduates, I did some research and we found that maybe $1^1/_2$ to 2 percent of all births do not fall strictly within the tight definition of all-male or all-female, even if the child looks that way. Beyond having a mixed set of genitals, you could have an individual with an extra Y chromosome.

He'd still look like a standard male, but he'd have this extra chromosome. Or you could have someone who was XO, a female with underdeveloped ovaries, known medically as having Turner's Syndrome.

My point is that there's greater human variation than supposed. My political point is that we can afford to lighten up about what it means to be male or female. We should definitely lighten up on those who fall in between because there are a lot of them.

— *You want a halt to sexual assignment surgeries on infants. Why?*

People deserve to have a choice about something as important as that. Infants can't make choices. And the doctors often guess wrong. They might say, "We think this infant should be a female because the sexual organ it has is small." Then, they go and remove the penis and the testes. Years later, the kid says, "I'm a boy, and that's what I want to be, and I don't want to take estrogen, and by the way, give me back my penis."

I feel we should let the kids tell us what they think is right once they are old enough to know. Till then, parents can talk to the kids in a way that gives them permission to be different, they can give the child a gender neutral name, they can do a provisional gender assignment. Of course, there are some cases where infants are born with life-threatening malformations. In those rare situations, surgery is called for.

— *In* Sexing the Body, *you suggest that estrogen and testosterone should not be termed sex hormones. You'd prefer we called them growth hormones. Why?*

The molecules we call sex hormones affect our liver, our muscles, our bones—virtually every tissue in the body. In addition to their roles in our reproductive system, they affect growth and development throughout life. So to think of them as growth hormones, which they are, is to stop worrying that men have a lot of testosterone and women, estrogen.

— *Among gay people, there is a tendency to embrace a genetic explanation of homosexuality. Why is that?*

It's a popular idea with gay men. Less so with gay women. That may be because the genesis of homosexuality appears to be different for men than women. I think gay men also face a particularly difficult psychological situation because they are seen as embracing something hated in our culture—the feminine—and so they'd better come up with a good reason for what they are doing.

Gay women, on the other hand, are seen as, rightly or wrongly, embracing something our culture values highly—masculinity. Now that whole analysis that gay men are feminine and gay women are masculine, is itself open to big question, but it provides a cop-out and an area of relief. You know, "It's not my fault, you have to love me anyway."

It provides the disapproving relatives with an excuse: "It's not my fault, I didn't raise 'em wrong." It provides a legal argument that is, at the moment, actually having some sway in court. For me, it's a very shaky place. It's bad science and bad politics. It seems to me that the way we consider homosexuality in our culture is an ethical and a moral question.

The biology here is poorly understood. The best controlled studies performed to measure genetic contributions to homosexuality say that 50 percent of what goes into making a person homosexual is genetic. That means 50 percent is not. And while everyone is very excited about genes, we are clueless about the equally important nongenetic contributions.

— *Why do you suppose lesbians have been less accepting than gay men about genetics as the explanation for homosexuality?*

I think most lesbians have more of a sense of the cultural component in making us who we are. If you look at many lesbians' life histories, you will often find extensive heterosexual experiences. They often feel they've made a choice. I also think lesbians face something that males don't: At the end of the day, they still have to be women in a world run by men. All of that makes them very conscious of complexity.

— *How much of your thinking about sexual plasticity comes from your own life? You've been married. You are now in a committed relationship with the playwright Paula Vogel.*

My interest in gender issues preceded my own life changes. When I first got involved in feminism, I was married. The gender issues did to me what they did to lots of women in the 1970s—they infuriated me. My poor husband, who was a very decent guy, tried as hard as he could to be sympathetic. But he was shut out of what I was doing. The women's movement opened up the feminine in a way that was new to me, and so my involvement made possible my becoming a lesbian. My ex and I are still friends. He's remarried.

— *So the antifeminists are right: women's liberation is the first step toward lesbianism?*

(Laughs) It's true. I call myself a lesbian now because that is the life I am living, and I think it is something you should own up to. At the moment, I am in a happy relationship and I don't ever imagine changing it. Still, I don't think loving a man is unimaginable.

— *What do you think nature is telling us by making intersexuals?*

That nature is not an ideal state. It is filled with imperfections and developmental variation. We have all these Aristotelian categories of male and female. Nature doesn't have them. Nature creates a whole lot of different forms.

January 2, 2000

Postscript

Anne Fausto-Sterling reports in an e-mail that she's "following up on ideas proposed at the end of my last book. So eventually, there will be a sequel."

We look forward to *Sexing the Body—Part Two.*

CYBERNETICISTS

MARVIN L. MINSKY

Why Isn't Artificial Intelligence More Like the Real Thing?

Rick Friedman

Marvin Minsky, 71, a rumpled-looking man who wears shirts mended with masking tape, is Toshiba Professor of Media Arts and Sciences and professor of electrical engineering and computer sciences at M.I.T., and one of the world's leading theorists of artificial intelligence. In the late 1950s, Dr. Minsky and John McCarthy, a professor of computer science at Stanford University, founded a research program that would evolve into the M.I.T. Artificial Intelligence Laboratory. In addition to inventing and building thinking machines, Dr. Minsky wrote the classic *The Society of Mind,* (Simon & Schuster, 1986), in which he tries to show how intelligence works "by the particular way the agents in the brain have evolved to interact." *The Turing Option* (Warner Books, 1992) a novel by Dr. Minsky and Harry Harrison, is about superintelligence in not-too-far-off 2023.

___ *In the 1960s, science students, particularly those at M.I.T., talked of artificial intelligence, or A.I., as if it would create world revolution. Were they too optimistic?*

Well, it got stuck. A.I. was able to produce all kinds of wonderful things ... Programs that did better than the average stockbroker or portfolio manager, programs that could fix some piece of equipment. Around 1980, progress stopped in some ways and people went off in a number of other directions to try to find some way to get back. It stopped because we'd done the easy things. In the eye of eternity, it got stuck for a moment.

A good example is that in 1964 or 1965 one of our students, Daniel Bobrow (now a vice president at the Xerox Corporation), wrote a program that could read a question from a high school algebra book, and sometimes solve the problem. So it could figure out a little bit of language and algebra. It didn't get most of the problems because it couldn't understand the words. What people tried to do then was get a program that would read a story from a first- or second-grade children's book. But what happened

97

was this: For any particular story, we could build into the program the knowledge necessary to read that story. We didn't have much trouble with the grammar. As soon as something was mentioned that the program didn't know about . . . (the system broke down). One M.I.T. student had a story where some person's daughter was kidnapped by the Mafia and they demanded a ransom. So he asked the program "What should we do?" The program couldn't understand. Finally, it asked, "Why would he pay MONEY to get his daughter back?"

It could figure out a little bit of language, a little bit of algebra. It didn't get most of the problems because it couldn't understand the words. As far as I know, nobody has been able to get a machine to solve real problems that are informally expressed, the way somebody would normally express them.

— *How do you define common sense?*

Common sense is knowing maybe 30 or 50 million things about the world and having them represented so that when something happens, you can make analogies with others. If you have common sense, you don't classify the things literally; you store them by what they are useful for or what they remind us of. For instance, I can see that suitcase (over there in a corner) as something to stand on to change a light bulb as opposed to something to carry things in.

— *Could you get machines to the point where they can deal with the intangibles of humanness?*

It's very tangible, what I'm talking about. For example, you can push something with a stick, but you can't pull it. You can pull something with a string, but you can't push it. That's common sense. And no computer knows it. Right now, I'm writing a book, a sequel to *The Society of Mind,* and I am looking at some of this. What is pain? What is common sense? What is falling in love?

— *What is love?*

Well, what are emotions? Emotions are big switches, and there are hundreds of these. . . . If you look at a book about the brain, the brain just looks like switches. . . . You can think of the brain as a big supermarket of goodies that you can use for different purposes. Falling in love is turning on some 20 or 30 or these and turning a lot of the others off. It's some particular arrangement. To understand it, one has to get some theory of what are the resources in the brain, what kind of arrangements are compatible and what happens when you turn several on and they get into conflict. Being angry is another collection of switches. In this book, I'm trying to give examples of how these things work.

— *In the 1968 Stanley Kubrick film* 2001: A Space Odyssey, *a computer named HAL developed a lethal jealousy of his space companion, a human astronaut. How far are we away from a jealous machine?*

We could be five minutes from it, but it would be so stupid that we couldn't tell. Though HAL is fiction, why shouldn't he be jealous? There's an argument between my friend John McCarthy and me because he thinks you could make smart machines that don't have any humanlike emotions. But I think you're going to have to go to great lengths to prevent them from having some acquisitiveness and the need to control things. Because to solve a problem, you have to have the resources and if there are limited resources . . .

— *Where were Stanley Kubrick and his co-author, Sir Arthur C. Clarke, right with their* 2001: A Space Odyssey *predictions?*

On just about everything except for the date. It's quite a remarkable piece.

— *Do you believe the NASA wastes money by insisting on humans for space exploration?*

It's not that they waste money. It's that they waste ALL the money.

— *If you were heading NASA, how would you run it?*

I would have a space station, but it would be unmanned. And we would throw some robots up there that are not intelligent but just controlled through teleoperators and you could sort of feel what's doing. Then, we could build telescopes and all sorts of things and perhaps explore the moon and Mars by remote control. Nobody's thought of much use for space. The clearest use is building enormous telescopes to see the rest of the universe.

— *Why are manned shots a NASA priority?*

Because NASA's people are basically oriented toward keeping themselves alive. They are a big organization. And the biggest part of it is Houston and that has to be fed, and what Houston is good at is putting men in space. The Jet Propulsion Lab is much smaller and has a smaller staff and is good at doing everything else. So, I think, in order to support that, they get into this vicious circle where you have to convince yourself that's what the public wants. Now, I think, the public is more excited by *Sojourner* than by astronauts.

— *When you go to the movies, what do you see?*

Terminator, Total Recall, which had ideas about implanted memory. Pretty clumsy, but I loved the engineering. I don't like movies exactly. One of my rules is not to think of the whole thing as having any unity. The idea of liking a whole movie is . . . People have this idea that they have to like something or not.

— *What do you read?*

Science fiction.

— *Do you read science fiction in the same way spies read spy novels—for ideas?*

Yes. There are a dozen very, very rich sources of ideas out there. Gregory Benford of U.C. Irvine, David Brin, Larry Niven are the best writers of our period. When they write a book, there's some big new idea about something. I've also gotten a lot of good ideas from old-timers like Robert Heimlein and the late Isaac Asimov.

— *Where was Mary Shelley right and where was she wrong with her* Frankenstein *last century?*

She certainly was right in predicting how people would not understand the poor thing. That's SUCH a sad story! By the way, I've gone through that book very carefully to see if she left any hints explaining how the robot worked. But alas, no clues and the funny part is when you read it, you don't mind.
July 28, 1999

Postscript

In addition to his regular work at M.I.T.'s Media Lab, Dr. Minsky is a Walt Disney Research and Development Fellow, and is—"currently working on a new book, *The Emotion Machine,* describing the roles played by feelings, goals, emotions, and conscious thoughts in terms of processes that motivate and regulate the activities within our personal societies of mind."

ANNE FOERST

Do Androids Dream?
M.I.T. Working on It

Rick Friedman

Cambridge, Massachusetts

Anne Foerst, 34, a researcher at the Artificial Intelligence Laboratory at the M.I.T. and the director of M.I.T.'s God and Computers project, apologized on a recent afternoon that a certain robot named Kismet wouldn't be joining our interview.

"Cynthia Breazeal, who built Kismet, is away in Japan right now and there's no getting her going," Dr. Foerst said in her German accent, "but you'd love her. She's oh so cute." A cute robot? Well, yes. At the Artificial Intelligence Laboratory, engineers are trying to build robots with social skills and humanlike experiences, and so, as an experiment, they've created creatures that they think humans will relate to.

Dr. Foerst, a Lutheran minister who supported herself by repairing computers during eight years of higher education in Germany, serves as theological adviser to the scientists building Kismet and the robot's brother, Cog.

—— *What exactly do people do here at this laboratory?*

We are trying to build robots that are social and embodied.

We have four projects. I am the theological adviser for two of them: the building of the humanoid machines, Cog and Kismet.

Cog is a robot built in analogy to a human infant. He has a torso, two arms, a head, ears, and eyes. He, it, learns to coordinate those limbs to explore its environment, just as newborn babies do. Kismet is a robot who interacts with humans through her body posture and facial expressions. The aim of this project is to explore social interactions between humans and robots and also between the humans themselves.

—— *Why a theologian here in this particular laboratory?*

Two reasons. The first is when you build machines in analogy to humans, you make assumptions about humans. Theologians explore the

103

cultural and spiritual dimensions of that very question, What does it mean to be human? The idea is that as these robots are built, we can use the wisdom of religious studies to enlarge our understanding of humans, and thus what you build into the humanoid machines.

The other reason is that when we build social interactive robots that force people to treat them as if they were persons, tricky moral questions come up. For instance, Who are we, really? Are all our reactions actually developed in a very mechanistic, functionalist way? Or is there a dimension to social interaction that goes beyond that? What are ethics here? Why should I treat someone else like a human, with dignity, when it is just a mechanistic thing?

For instance, one question we discuss quite frequently is, What would be the threshold when the robots are developed to a certain point that you couldn't switch them off anymore? The question really is, When does a creature deserve to be treated as intrinsically valuable?

— *When do you think a robot should be treated as intrinsically valuable?*

Well, that moment is 50 years down the road. At least. But it's pretty clear that when it comes, those who built the robot will have to make that decision because they won't be blinded by their fears of the seemingly human qualities of the machines. They'll know what's inside. And if it ever got to the point where the builders felt, oops, now that has become something, the builders could become the creature's strongest advocates.

— *What make the robots Cog and Kismet different from previous ones?*

Previous attempts put very abstract features of human intelligence into a machine: chess playing, mathematical theorem proving, and natural language processing. The idea now is, In order for a machine to really be intelligent, it has to be embodied. We say intelligence cannot be abstracted from the body. We feel that the body—the way it moves, grows, digests food, gets older, all have an influence on how a person thinks. That's why we've built Cog and Kismet to have humanoid features.

Cog moves and experiences the world the way someone who can walk upright might. He experiences balance problems, friction problems, weight, gravity, all the stuff that we do, so that he can have a body feeling that is similar to ours. The humanoid features are also crafted into the machines in order to trigger social responses from the people interacting with them.

The other thing we believe is that humans are human because we are social. Thus, we try to treat Cog and Kismet something like the way most of us treat babies, as if they have intentionality, emotion, desires, and intelligence. We give them as much social interaction as we can.

Cog is a whole body and Kismet is mostly a head and facial expression. Our work with Cog concentrates more on the embodiment stuff and Kismet more on emotional-social learning.

— *Is the robot Kismet a she?*

Robots are its. But I can't help but think of her as a she. If you were to see Kismet, you would be taken by her enormously expressive face: long eyelashes, big blue eyes, movable brow, cute, kissy mouth. When Kismet puts her eyes on you and looks sad, you want to make her happy. Of course, part of you thinks, It's just a stupid machine. But you do react and you can't help it.

The point of reacting to Kismet is the same as reacting to a baby. We believe that only when you treat the machines as if they have all these social characteristics, will they ever get them. If you want to have an intelligent being, you need to create that circle. So we react here to Kismet's emotional displays. When she's bored, you want to make her happy. When she seems scared, you back off.

— *Has the very social robot Kismet done anything yet that has astonished you?*

Kismet has not yet learned. Cog is the one who learns. A former graduate student, Matt Williamson, the guy who built Cog's arms, taught the robot how to control his arms.

To coordinate the arms, Matt had to touch a part of Cog's body and then, the arm would touch that part, too. After he did that for the first time, Matt ran into my office and said, "You've got to come to look at this." It looked so eerily human. It's not so much that Cog does something that's unexpected, it's more the human reaction, like, it's alive!

— *People often talk about humans having some indefinable extra above life that makes for humanness—some call it "spirit." Can a robot have spirit?*

Rod calls it "juice." He says, "Even though I get it all right, might there not be some juice I'm missing?" I would say from a religious perspective, the juice is that which comes from the outside world and emerges in social interaction.

— *Some people might complain that in building humanoid robots, you are trying to supplant God.*

Yes. I know. They say, "Do you want to be like God?" Actually, if you use biology as your inspiration in your robot-building and focus on embodiment and environment, you get much more humble instead of arrogant. Suddenly, you realize that even the most brilliant robot that the most brilliant engineers have worked on for years and years is still dumber than an insect.

— *So, in your view, God is, as the Latin Americans say, the "intellectual author" of everything?*

No. The creative author. When we are creative, the power of creation is from God.

— *In the many plays, novels, and movies about robots, the dramatic climax of the story always comes at moment when the machine achieves sentience. Why do you think that is?*

Well, I think it's the search to feel and to be treated like something more than the sum of the parts that's inherently dramatic. This is the moment when the robots start to participate in the all-too-human quest of what does it mean to be me?

— *In the movie* 2001: A Space Odyssey, *HAL becomes a danger to humans once he's sentient.*

In *Frankenstein*, too. But in both cases, there is an explanation. When you look at Frankenstein's robot, he is never part of a community. His creator left him right away. The people hated him, feared him, ran away from him. The only person who ever loved him was a blind man who couldn't see what he looked like. The robot was never treated as a valuable being, a person with dignity. He had to turn against the society that shunned him. Where should the goodness come from when he never experienced it himself?

HAL is the same thing. And he's disembodied. There is no body with which to experience the world. I would even say that in such a setting a robot couldn't even become sentient. In the movie, HAL becomes sentient at some point and nobody notices. No one treats him properly and he's isolated and what happens? He becomes psychotic.

— *What's your favorite robot movie?*

Blade Runner. I teach it in my classes. The robots have this absolute search for meaning, and when their quest is not taken seriously, it becomes fatal. The movie raises this wonderful question: How do humanoid creatures feel about having been created by us and how do they deal with their human-made limitations?
November 7, 2000

Postscript

Anne Foerst's life went into a tailspin in the months directly after our interview was published. First, her marriage—not yet three years long—ended.

Suddenly, Dr. Foerst was faced with big life- and career-changing decisions to make. At M.I.T., the powers-that-be didn't seem ready to offer her a professorship, she told me in a phone interview in March of 2001.

And so, with love and work up for grabs, Anne Foerst rebooted her life at a Franciscan university in upstate New York, St. Bonaventure's, where she now is a professor of computer science and theology and where she's charged with setting up a center that investigates the places where the two disciplines might have something to say to each other.

"I love the symbolizism of working at a university called 'St. Bonaventure,'" Foerst enthused in her German-accented English. "The great thing about St. Bonaventure, the Saint, was that he was one of the big opponents of Aquinas and the thing that he rejected most was the mind-body split and the concentration on rationality."

After the interview was published, Dr. Foerst heard from a top New York literary agent who took her on as a client. Together they are working on soon presenting a manuscript to commercial publishers that will be called *On Robots and Humans and God.*

MICHAEL L. DERTOUZOS

A Pragmatist on What Computers Can Do

Rick Friedman

In eight books, **Michael L. Dertouzos**—engineer, inventor, theoretician, and director of the Laboratory for Computer Science at the Massachusetts Institute of Technology—predicted the many ways the information revolution would affect human lives.

In the 1970s, he forecast that one third of all American homes would soon have personal computers. In 1980, he announced the coming of a worldwide Internet culture.

So what is Dr. Dertouzos thinking about now? Over cups of hot tea, in his Cambridge office, he offers some hints.

___ *You recently gave a deposition in the Microsoft antitrust suit. How does your old friend Bill Gates feel about the testimony you offered?*

I suspect he may be a little disappointed. I was asked in my testimony, "Are browsers part of the operating system or applications?"

"Today, browsers are applications," I answered. "Tomorrow, they will be merged." That was the point where Bill and I didn't see eye to eye. The Microsoft position is they've already merged.

___ *Was it hard for you to decide to testify in the Microsoft case? M.I.T. has received a $20 million gift from Bill and Melinda Gates. He's also a friend of yours.*

When the Microsoft lawyers first called me, I said no. And then Bill said, "All I want you to do is testify on what you have written." So then I said, "All right, Bill, no money in expert witness fees, no prepping by Microsoft lawyers, and I will only testify on the merger of the browsers and operating systems." I gave a six-hour deposition. After that, Microsoft dropped me as a witness.

___ *When you gave that deposition, you also said that you differed with Gates on the effect of the computer revolution on the rich and the poor. What did you mean?*

109

We differ on that and on some of his views on "frictionless capitalism." He thinks consumers and suppliers are going to meet on this gigantic football field called the Internet and they are going to do deals together without an intermediary. It's a seductive idea.

In my opinion, it is right for about 15 percent of the marketplace. But wrong for 85 percent. It will happen on standard products and products that do not involve trust questions—relatively small products.

Bill sees this expanding world of networking as an opportunity for poor people to sell their wares, get educated, participate in the world marketplace and pull themselves up from poverty. I see the exact same thing with a time scale of 15 years—and only if we help.

— *How did you arrive at a 15-year time frame?*

I learned it from Nepal. A while ago, I had this naive assumption that I could go to Nepal, obtain computers and training for the Nepalese and get them to have a 20 percent jolt in the G.N.P.

But here's what I found out: Only 30 percent of the Nepalese are literate. Of that 30 percent, only 10 percent speak English. Even if I got someone to provide every one of them with a computer with communications, what could they do with them? They have no skills to sell.

To get people to do this, I would have to educate them, and people don't get educated overnight. So, 15 years. From this and other experiences, I've concluded that the information revolution, if left to its own devices, will mean that the rich are going to buy more computers, be more productive and become richer, and the poor will not be able to do that and will stand still.

History teaches us that whenever the gap between rich and poor increases, we have all kinds of troubles.

— *On a slightly different subject, "wire up the schools" has become the latest solve-all slogan that politicians offer in their campaigns. Does it seem a little mindless to you?*

Yes. Last year I was at a conference with Benjamin Netanyahu, then Prime Minister of Israel, and he announced: "I want to connect all toddlers in my country. I have 300,000 toddlers under 5 and I want to connect them, but I'm having trouble finding the money." I said, "Mr. Netanyahu, why do you want to do that?" He said, "Isn't it obvious?" My position is: let's experiment, let's connect a few thousand kids, let's not deploy this for tens of millions because the jury is still out.

It is good for some things—like training people in skiing and tank-driving. But the bottom line is that we in America have all the computers in the universe and we still rank 18th in math and physics worldwide.

— What have you noticed about the way politicians seize on issues dealing with technology?

That they haven't matured yet. The trend is still, "This is modern—it is evidently good." They don't worry about other things concerning technology—for instance, privacy issues.

For instance, in the United States, every time you do a transaction, you are asked for information and you have no control over what will happen to it. You are, in fact, commoditizing your privacy in return for the instant gratification of getting some goods. It means that a lot of things about you will be known. I'd like to see legislators debating these issues instead of grandstanding and saying "wire all schools."

— Speaking of Internet shopping, have you used eBay?

Yes, I've played at bidding. I was experimenting. I wanted to see the elasticity of the bidding. I just pushed a poor woman up on her bid on a rare Tennyson book. I bid it up another hundred dollars and pulled out at the last minute. It wasn't a nice thing to do, but I was trying to see how responsive the medium is.

— There's been a nut fringe that went crazy at the millennium. Why is it this time so focused on technology with Y2K?

You gave your answer by asking your question. It is in the nature of human beings—to go crazy about potentially apocalyptic events that have no basis in rationality. Now, Y2K is a fully rational bug. We know it's going to be trouble. So now we have the apocalyptic tendency of human beings, with a real bug that could hurt a lot of things. I take the position that, yes, we are going to have a few bad cases, but the bulk of it is going to be mostly a nuisance.

The minute I took this position, I heard from a large number of people who said, "How can you say this—the world is coming to an end!"

— *If you had the power to redesign the curriculum of an engineering school, how would you change it?*

I would make it a lot more sensitive to the whole human being. So I would study more some of the seminal key aspects of human activity— historical, political, psychological. For example, I would take key issues that we're facing today, for instance privacy, and say, "Where else have we had problems in privacy before?" I would take a situation in technology in which loyalty comes in conflict with the truth and see how it plays out. Should you tell the truth or be loyal to a friend or hide it?

— *Was that the dilemma you faced when you gave a deposition in the Microsoft case?*

No, I wasn't torn at all. Nobody is going to make me say something that I don't think is true. So I was really never torn. I do value money and loyalty, but in any business the biggest capital we have is sticking to our guns.

— *You grew up in Athens, in the 1940s and 50s, where your father was an admiral in the Greek Navy. During those years, did you ever dream you'd become a priest at this temple of high tech?*

No, but I went to a Greek-American school. So I was imbued with high tech. By the time I was 13, I was reading the books of people who work at this institution. I was riding bicycles with sails on them. And I built a radio station. I absolutely related to America. I believe I'm a mixture of both worlds, and I'm absolutely comfortable in both.
July 6, 1999

Postscript

Dr. Dertouzos published his latest book, *The Unfinished Revolution: Human-Centered Computers and What They Can Do for Us,* from HarperCollins in February of 2001.

He is seeking to "finish" the cybernetic revolution through his M.I.T. Oxygen Project, which is described on the M.I.T. Web site as aiming toward a future "where computation will be freely available everywhere, like batteries and power sockets, or oxygen in the air we breathe. Computation will enter the human world, handling our goals and needs. We will not need to carry personalized devices around with us. Instead, 'anonymous' devices will personalize themselves in our presence by finding whatever information or software we need."

According to a recent e-mail from Michael Dertouzos, "The Oxygen Project to build such a prototype at M.I.T. is on its way with $50 million and five years to go."

JOHN MAEDA

When M.I.T. Artist Shouts, His "Painting" Listens

Rick Friedman

Within the art world, **John Maeda**, 32, is an anomaly—a prize-winning graphic designer and kinetic artist with a fistful of engineering degrees from the Massachusetts Institute of Technology.

From his base in M.I.T.'s Media Laboratory, where he is the Sony Career Development Professor of Media Arts and Sciences and directs the Aesthetics and Computation Group, Professor Maeda uses the computer as a tool and medium to create art that can be produced only digitally and that has the specific look of the new technology.

One of his best-known pieces is a drawing called *Time Paint*, in which colors fly through space. Another piece, *The Reactive Square*, is about squares that change shape when a viewer shouts at them.

Professor Maeda's computer-generated art pieces have been shown at the Ginza Gallery in Tokyo and in that great digital art gallery—the World Wide Web. We talk in his dark computer-filled Cambridge offices on an otherwise sunny summer day.

— *Your last book,* Design by Numbers *(M.I.T. Press), is an art book that is also a manual for a new computer language that you invented to help artists understand the guts of computer design. Why create a whole new computer language?*

Several reasons. I wanted to make a system that was free of charge. Second, all the programming languages are in English. Because this one is free and distributed on the Web, we are now offering it in French, Spanish, and German. It will soon be in Korean and Japanese.

The third reason was that programming languages are made for people to write programs—big applications.

For someone just starting out making art on their computers, they don't want this big Mack Truck of a system. They just want a simple bicycle that they understand.

So I designed the visual equivalent of a simple bicycle. *Design by Numbers*, D.B.N., was an attempt to demystify the technology behind computer art, to show how simple it is, and that people can do it. D.B.N. can run directly within any Java-enabled Web browser. The Web site address, by the way, is dbn.media.mit.edu.

___ *When you are creating your own computer art pieces, do you ever use prepackaged drawing programs like Adobe?*

Oh, yes, all the time. There are all kinds of fine touches that prepackaged software makes easy. I could invent my own finishing system, but this is faster. Of course, the basic ideas, I create.

___ *Yet, I sense you're scornful of Adobe users?*

The problem is that most people can't just "finish" things with it. They have to use it to start them, also. For much of recent history, people have created with brush, ink, paper—the materials of art. Now that they have begun creating with software and computers, the styles that emerge are homogeneous because the software is universal.

Without being able to know how to program, you can't break out of the technology—just like if you don't know how to use brush and ink, you're limited.

For most people, this really isn't a problem; they aren't necessarily looking for anything new. But for people who are seeking the next step, the prepackaged becomes an impossible barrier to break free from.

I make everything I do. Many people are surprised that I don't have a programmer making things for me. And others are surprised that I don't have an artist controlling me, telling me how to program. Because today, people don't realize that it is possible to think and create on the computer. Artists are used to thinking that programming is very hard— impossible.

And technologists are used to thinking that they can never become artists. Me, I just make things. It's just a natural flow of action and thought. If people see, "Oh, he does that," then maybe they'll think, "I can do it too."

___ *But lots of nonartists use computers for creating images . . .*

They are using it as a tool, but not as a material. And to use it as a tool, you need to understand the medium, which means understanding the technology. Young people are changing this because they have grown up with computers.

— *If a conventional artist produces an object on a computer, does that automatically make it art?*

It's art, but it's just a painting and no different from conventional art. It's not intrinsically different or superior just because it was created digitally, and it's not digital art. Because digital art starts with an understanding and appreciation of the medium which, unfortunately, is today programming.

— *When you were growing up in Seattle, what were your aspirations?*

My aspirations were my father's—and his were that all of us kids would go to either Harvard or M.I.T. He didn't know what or where they were.

He was a cook and a tofu maker by trade. These were the only two schools he'd heard of. His were immigrants' dreams.

Also, I was very good at math and at drawing and I had a teacher who said, "You should go to M.I.T. because it's a good place for math."

— *Did the term "artist" ever come on your screen as a child?*

Some of my teachers felt that I should become an artist, but my parents looked down upon this because they pictured me as poor.

So I got to M.I.T. in 1984, a time when, all of a sudden, the computer had become visual. There was a great demand on the campus for people who could design icons, choose fonts. I worked at different places on campus doing this type of work and I thought I was so good at it. People used to tell me, "maybe you should be a graphic designer." I didn't know what that was. I only knew what engineers, tofu makers, and Japanese restaurant owners did.

Then, I saw this book, *A Designer's Art*, by Paul Rand, and I was shocked. Paul Rand did the IBM, ABC, UPS logos, and I saw how exquisite his power over space was, and I realized I wanted to do this. I told my father. He said, "No way. You must be an engineer." So I went on and got my bachelor's and my master's in engineering, and once I did that, I told my father that I still wanted to be an artist and he said, "Now that you're a man, you can do what you want." So I thought the very place to go was the M.I.T. Media Lab and I joined the Ph.D. program here.

— *Were you happy at the Media Lab?*

Not happy, on two levels. The Media Lab was more a technology place than an art place, and I was constantly reminded of that fact. I was fortunate to know Muriel Cooper, the design professor at the lab, and she gave me the best advice, which was, "Go away." So I went away. I went to Tsukuba University, the great classic design school outside of Tokyo.

— *What did studying in Japan teach you?*

The most important thing was not to be embarrassed about who I was. I had always been embarrassed about coming from a manual-labor family. In Japan, I was studying conventional art, and I used my hands all the time. That made me feel in touch with my human side, which I had lost when I came to M.I.T. I think the people who make things in Japan are kind of elevated. There are television shows there where cooks and noodle makers are shown as heroes. The ordinary, well done, is celebrated. So, once I realized that I came from a tradition that mattered to me, I gave up on technology and swore it off.

Then, one day, my teacher at the art school asked, "John, what are you planning on doing with your life?" I said, "Honor the classics, sir."

He accused me of being the Japanese equivalent of a horse's rear end.

"When you're young, you should do things that are new. When you're old you can always go back to the classics. They'll always be there," he said.

From that point, I went back to technology and rediscovered everything. I became a sort of born-again technologist.

— *Does the new technology mean the end of art as we know it?*

Yes, it does. It represents a new dimension to the way art will be understood or perceived.

It's a departure from appreciating a singular moment. What that means is . . . the reason why we can appreciate art is because most art has a single resting point: canvas.

It's painted. It's dried. It aspires to be perfect. The medium of the computer is continually shifting. It can shift at will, in a microsecond. Or an hour.

There's no limit on how it can be taught to change.

— In the 1960s, there was a kind of avant-garde-technoart movement that came out of Bell Labs. Do you like the abstract technoart of the 60s?

Definitely. Because they were pushing the limits of their medium. They had a vision, but they didn't know how to make it succeed.

I like a computer artist named Michael Noll, who was a Bell Labs scientist. He could make all these things, but he didn't know what they were and didn't know how to contextualize them. Karl Gerstner, whom I respect the most, was a visionary artist/designer, who foresaw the potentials of computers, but didn't know how to realize them.

— Does this mean that the artists of the future will all be going to M.I.T.?

I think artists of the future will have to go to better art schools, which will teach them how to control technology instead of being controlled by it. M.I.T. is one option. One of the things that brought me back to M.I.T. was the idea that engineering will be the humanities of the twenty-first century. I believe there is room in this world for a humanist technologist.
July 27, 1999

Postscript

Interviewing John Maeda was _Science Times_ Editor Cory Dean's idea. She'd met him at a conference and was impressed. "Just ask him to tell you his life story," she advised.

And I did. And, of course, it was fascinating. Maeda is as good a storyteller with the narrative of his life as he is a visionary with images. This is one of my favorite interviews. Ever. The communication was good.

"I became the Associate Director of the M.I.T. Media Laboratory in September of 2000—in addition to my usual roles as associate professor of design and computation at that lab," he wrote in February 2001.

Mr. Maeda has also had art exhibitions all over the country, including one at the Christinerose Gallary in New York that he said came to him because of the _Science Times_ interview.

COMMUNICATORS

SHAWN CARLSON

Just Like a Film Script, From Jobless to Genius

Rick Friedman

Providence, Rhode Island

Shawn Carlson's biography reads like a mutant graft of Mr. Wizard meets Horatio Alger: An unhappy physicist working in a mainstream laboratory decides to quit his job and start a nonprofit organization aimed at encouraging the projects of backyard tinkerers and garage experimenters.

As Dr. Carlson devotes himself to organizing his Society for Amateur Scientists, he drives his family to near-penury. Just as he is about to go broke, the John D. and Catherine T. MacArthur Foundation comes to the rescue with a "genius grant," and nearly $300,000.

"After MacArthur, I was a certified 'genius'; the day before I was just certifiable," Dr. Carlson, 40, said over coffee one morning in Providence, where he and his family have recently moved. Stocky, with sandy-blond hair, he has an open face and an easy laugh.

In addition to his organizational work, Dr. Carlson has written the "Amateur Scientist" column in *Scientific American* magazine for the past five and a half years.

The column, however, was suspended in March of 2001 to make room for "other good ideas for columns," said John Rennie, the magazine's editor in chief. "Amateur Scientist" had been a regular feature of the periodical since 1928.

— *Define "amateur scientist."*

We think of amateur scientists as anyone who is engaged in scientific inquiry for the sheer joy of it. If you include all the home tinkerers and amateur naturalists, there must be several million in the country.

— *How American is the movement?*

As American as apple pie. Many of the Founding Fathers were amateur scientists—Benjamin Franklin.

Since the mid-1800s, farmers and everyday folk have been organized to make weather observations with the Smithsonian Institution. Thomas Jefferson organized weather watchers in Virginia and gave them instrumentation to study climate change.

— *Aside from the occasional eclectic discovery, are there areas where amateurs can make a discernible difference to the general body of scientific knowledge?*

Most scientists don't really realize the tremendous contributions that amateurs have made over the decades. For instance, most of the global warming debate hinges on data collected by amateurs before the 1950s— farmers, people like that. Bird watchers have done incredible work on calculating the fate of various species. On environmental questions, I see a situation where amateurs could become a vast data collection army, measuring all kinds of things—thickness of birds' eggs, magnetic field fluctuations.

— *What exactly does the Society for Amateur Scientists do?*

Democratize science. We hope we help encourage the scientific creativity in the everyday citizen. We believe that you don't just have to be a Ph.D. to do interesting science. S.A.S. serves as a switchboard for amateurs to trade ideas. In addition to publishing our bulletin, we print how-to information for projects and provide access to inexpensive equipment and chemicals. We also have mentors to help with projects, and a Web site: sas.org.

The Amateur Society has just opened up an observatory with a 16-inch telescope inside New Milford High School in Connecticut. Now the question is, What is the mission going to be? We're still having meetings about that.

First, we want to get the students in there, but there is going to be a tremendous lot of open time available on the thing. I'd love to link this to the Internet, so that amateurs around the world could use it, for a small fee. The idea is, If we can get the amateur astronomy community on it, we could maybe make $1,000 a night and then have the money to constantly upgrade the telescope.

— *Name a few amateur projects that have come across your desk and that have intrigued you.*

There's a guy named Roger Baker, and he has a design for a magnetometer that allows you to measure changes in the earth's magnetic fields. He uses materials bought at Radio Shack for about $10.

A wonderful project was done by two dear friends: George Schmermund and Greg Schmidt. They have developed a system to measure extremely tiny masses with very high precision.

The secret is to go to a surplus store and to buy one of those old analog meters. George invented a way of taking an old discarded galvanometer and turning it into a device that can measure a millionth of a gram. And then, my friend Greg computerized that device for a couple of dollars and turned it into a really powerful instrument so that now any amateur can measure tens of micrograms.

— *Tell us about your grandfather, the late George Donald Graham.*

Grandpa Don was a free-spirited wild man, a person with incredible scientific creativity. But he was one of those personalities who couldn't go through the standard course of instruction. So he pursued his passion for mathematics and geology and biology on his own and he would frequently write really interesting papers that he couldn't get published because he didn't have "Ph.D." after his name.

My grandfather would produce a brilliant paper on gravity and corn—and it just killed me that someone of this kind of talent would be excluded from the discussion just because he didn't have the credentials. Grandfather wore these rejections like a badge of honor on his sleeve. The only place that would publish him was the "Amateur Scientist" column of the *Scientific American,* a column which I authored for five and a half years.

— *How much are you your grandfather's boy?*

I'm very much like him. I've got his scientific talents. But as a younger man, I didn't want to make his mistakes—so I got myself a Ph.D. in nuclear physics from U.C.L.A. and worked as an astrophysicist at Lawrence Berkeley Laboratory from 1991 to 1993. There, I discovered I was like him in another way: I couldn't work on someone else's project, and I couldn't keep a job. I drifted more and more toward involvement with the amateur scientific community. Then in 1994, my wife and I left the Bay Area. She was about to enter graduate school in San Diego. I decided to use this change and take a huge chance to try to make a contribution as an amateur scientist.

I took the little savings we had and put it toward starting an organization for people like me—a society for amateur scientists. I had dreamed about doing this for 10 years. It was a way of paying tribute to my grandfather.

___ *How does an unemployed astrophysicist get to be an organizational entrepreneur?*

It was hard. I never took a salary. Mostly, my wife, Michelle Tetreault, and I ran up credit card debts and put our own money into it. We had a child, eventually two, and we lived on my wife's graduate student stipend. In 1995, I was asked to take over the "Amateur Scientist" column in the *Scientific American* and that provided $1,000 a month, later $1,500. You can imagine the difference that money made.

___ *Why were you so poor? Did your organization have difficulty finding funding?*

Oh, yeah. We fell outside the normal funding guidelines. We couldn't get research dollars because—guess what—our investigators didn't have Ph.D.'s after their names. We couldn't get educational money because we weren't focusing on children. On the other hand, we got a lot of moral support from prestigious science professionals. People like Kip Thorne, Douglas Osheroff, and Robert G. Coleman were our advisory board.

The financial problems all came to a head in June of 1999, when Michelle was about to finish up her Ph.D. and have our second child. One day, she sat me down and said, "You may have to close S.A.S. down because you have to take care of your family." And just like in the movies, as I stood there shaking my head, saying, "no, no, no," at that very minute, the phone rang with the news that the MacArthur people had granted me a "genius" fellowship—$290,000! Once the MacArthur prize hit, we were legitimate.

___ *Next month, the* Scientific American *will cease publication of the "Amateur Scientist" column—a feature that magazine has run with different authors since 1928. What are your feelings about the demise of a column that has meant so much to you?*

Well, of course, I'm heartbroken.

— *On a lighter subject, when you want to shop till you drop, where do you go?*

I would like to say Radio Shack, but I would have to say that my favorite place is Hyatt's Electronics in San Diego. It's this little hole in the wall run by this guy named Dick Hyatt. In there, you can find anything. I go and say, "Dick, I need 250 resistors." And he'll say, "Fine, I've got six million. What values do you want?"
January 23, 2001

Postscript

Shawn Carlson was a touch sad when last we spoke. It was during the Christmas holiday of 2000. At the *Scientific American*, a revamping of the magazine had eliminated his "Amateur Scientist" column, and he couldn't quite believe that something so important to his life was gone. The last time I spoke with Carlson, he was busily searching for a new home for the feature—a long-shot effort in a general magazine environment that is increasingly conventional.

An e-mail from Carlson in February of 2001 was more optimistic. Our interview had appeared, and he'd been deluged with offers from agents and book publishers—a pleasant turn of events. "Since my interview appeared," he wrote, "I can add that I'm working on a new book, called *The Amateur Scientists*. It's going to be *Profiles in Courage* meets *Longitude*, by telling the stories of triumphs achieved by modern day amateur scientists. I hope it will appeal to anyone who's fascinated with the hero's journey."

This "hero" had also found some money to develop a pilot program for a potential PBS series to be called, what else? *The Amateur Scientist*, starring—*tra-rah*—Shawn Carlson!

JOHN HORGAN

A Heretic Takes On the
Science of the Mind

Susan Farley

The genial freelance writer greeting me at the door of his rustic home in Garrison, New York, on a warm August morning is **John Horgan,** 46, the unofficial bad boy of science journalism.

In 1996, Mr. Horgan, then a senior writer with *Scientific American,* published *The End of Science: Facing the Limits of Knowledge in the Twilight of the Scientific Age,* a 281-page essay in which he argued that scientific inquiry has gone about as far as it can go and that the questions remaining are unanswerable. Many scientists were outraged, but the book sold nearly 200,000 copies.

Mr. Horgan will no doubt be making a new set of enemies with the release of his latest work, *The Undiscovered Mind—How the Human Mind Defies Replication, Medication and Explanation* (Free Press). "I think of myself as a heretic," he says, "who is challenging the central dogma that scientific progress is eternal."

— *Tell us how you got the idea for this new book.*

It's really a follow-up to *The End of Science.* There were criticisms of my first book that I thought didn't have much substance, but one I thought was reasonable was that the science of the human mind, of all areas of science, had the most potential to be really revolutionary. So I wanted to see how far we had gotten with not just neuroscience, but psychology, psychiatry, behavioral genetics, the new Darwinian social sciences, artificial intelligence.

And what I found is that despite a lot of hype and despite some amazing instruments—MRIs, PET scanners—we seem to be spinning our wheels. We are not learning the kinds of things we want to learn. We aren't learning how matter can create a mind. We aren't even doing something practical like understanding schizophrenia. Or coming up with better treatments for it. Or even a cure. The practical issues are what people care about. On that, I've concluded, despite all the hype about psychopharmacology, especially, there has been very little progress in understanding mental illness and treating it.

___ *Why do you think we're not making any progress in understanding the mind?*

I don't think that there's a mystical barrier. It's just—the brain and the mind are fantastically complex.

___ *Sir Martin Rees, the Astronomer Royal of Great Britain, once told me that the understanding of outer space is relatively simple—molecular biology was what was complex.*

Yeah, and the brain! So he's agreeing with me that cosmology and particle physics are all wrapped up?

___ *I don't think so. He's simply saying one is a harder nut to crack than the other.*

Oh, there's no question: particle physics is like a children's game compared to neuroscience. There are a handful of particles that behave according to fixed rules—if you control the situation enough, you'll always know how those particles are going to act. With humans, you never know!

___ *Some would say that unpredictability is part of the mystery of what makes people human.*

Except I don't think you have to resort to mysticism to talk about the limits of mind-related science. There's no mystical reason why we can't do these things. It's just turned out to be extremely difficult. There's no law of nature that says that just through sheer effort and will, scientists can solve every problem. People should at least consider the possibility that, in some respects, we might not solve this thing.

___ *Do you enjoy debunking ideas about scientific progress?*

I think that "scientific progress" is idolized by people who think they are too rational to believe in a Christian God, or some form of religion . . . there's an almost worshipful belief that this extraordinary period of technological and scientific progress is just permanent, that it will continue as long as we have the will. I believe that science itself tells us that there are going to be limits to this process and that those limits are appearing right now. People refuse to acknowledge those limits because they have this faith that it can't end.

— *Don't you think that the new antidepressant medications—Prozac, for instance—have been a big step forward?*

In my book, I present evidence suggesting that most of the effect of Prozac, if not all of it, is due to the placebo effect. What is remarkable is that lots of psychiatrists will admit this is true.

I have friends who swear that Prozac saved their lives. I try to be very careful in the way I talk about some of these treatments because I know that people believe in them strongly, and that the belief itself can heal people.

— *Oh, come on, you debunk psychopharmacology throughout the new book. Aren't you trying to have it both ways?*

(Laughs) Well, I am. In the book, I come up with this idea of hopeful skepticism. . . . I think it's important that people know that the clinical trials show that Prozac is no more effective than older antidepressants or, as a class, more effective than psychotherapy. Now, all these things are better than doing nothing . . . but that can also be said of cognitive behavioral therapy or your local astrologer.

— *Prozac had a brilliant popularizer.*

You mean Peter Kramer (author of *Listening to Prozac*)? It did. What's interesting about Mr. Kramer's book is that he has a lot of deniability built into it. He certainly doesn't say that Prozac helps everybody, but the effectiveness of his case studies is very powerful. It's essentially the same method Freud used to sell psychoanalysis with stories of Anna O. and the Wolf Man, compelling literature selling a method of treatment.

— *A writer in the mind sciences you seem to like is Oliver Sacks. Why?*

He's an anti-reductionist. If you look at the history of mind-related science, you see lots of abuses of reductionism—social Darwinism, eugenics, now psychopharmacology. With Dr. Sacks, the underlying message of all his case studies is that each person with a given disorder is absolutely unique and has to be understood on his or her own terms.

— *What's your critique of most science reporting?*

I have enormous respect for all my science-writing colleagues, but, in general, I'm distressed that science writers aren't more critical of science. We often don't get as much sophistication out of science writers as we do out of sports writers. Or political writers. Scientists are very good at intimidating science writers. They are always telling us you can't "really understand" science unless you are a scientist, which is absolutely absurd.

— *When you were a staff writer at* Scientific American, *did you sometimes feel frustrated?*

Actually, I was encouraged to write critical articles, but the fit became uncomfortable over time. I guess I became too critical. There were certainly some people who thought I was anti-science. There were some people at *Scientific American* who were horrified by my last book, by what it said and by the title. I was challenging the central dogma, this faith in scientific progress. I think it's fair to say that I left by mutual agreement.

— *What are you working on?*

Both my books have raised the possibility that the mystical experience can provide insights into the universe as a whole and our minds that can complement science. I brought it up in a rather teasing way, but that's what I'm doing my next book on. The problem with a topic like mysticism is that it is so broad that it very quickly gets out of hand. I'm talking to Buddhists, yoga practitioners, and also a community of people who use psychedelic drugs as a way of inducing these experiences. At this point, I'm all over the place.

— *Are you going to call the next book* The End of Reason?

No. Because I think of myself as a reasonable, rational person, and the challenge of this book will be to approach mysticism while hanging on to reason and objectivity.
September 21, 1999

Postscript

Zac, the yellow-collared macaw perched muse-like on John Horgan's shoulder for the photograph accompanying our interview, was one of two big budgies sqwarking around John Horgan's Garrison, New York home when I visited there in the summer of 1999.

Writers, like witches, need familiars around them while they work and Horgan, most originally, lived in a near-aviary. "My wife rescues parrots," he explained. "People get these expensive birds and then they find that they're a lot of work and they abandon them. " "Are parrots amusing pets?" I inquired.

"No, they are incredibly needy," Horgan replied, though he didn't elaborate on what he meant.

To this day, I retain a picture in my mind of John Horgan steaming away on the computer surrounded by this flock of ultra-demanding birds and dreaming no doubt of owning goldfish. Yet, somehow, this is a situation that works for him. John Horgan is one incredibly prolific writer.

And a provocative one. His books always manage to enrage someone. Alas, *The Undiscovered Mind,* the work I interviewed him about, did not prove as successful as his previous offering, *The End of Science.* But it did well enough—and it was widely reviewed.

In February of 2001, when I wrote Mr. Horgan for an update on his life, the following e-mail came back:

"Hi Claudia. I'm just finishing a book on mysticism for Houghton Mifflin, which I mentioned when we spoke. It considers whether mysticism and rationality are complementary or incompatible modes of knowledge. It should be published in a year or so. The current title is *The Deep End: Getting to the Bottom of Mysticsm.* Next I'm thinking of writing a book on how utopian ideas persist in our culture even though utopias are thought to have been discredited by Nazism and totalitarian communism. But if I don't get a decent advance for that, I'm going to become a daytrader."

IRA FLATOW

Latter-Day Mr. Wizard Expounds on the Joy of Science

Ira Flatow is one of the most influential communicators of science.

His National Public Radio talkfest, *Science Friday,* is a weekly dialogue with scientists and members of Mr. Flatow's audience of two and a half million, who phone in questions, ideas and opinions. And for six years, he was host and writer of *Newton's Apple,* a PBS science program for children.

But don't call him a popularizer. "You don't have to popularize science," he said recently over coffee in Manhattan. "All you have to do is make it accessible to people." Mr. Flatow grew up in Brooklyn and on Long Island. In 1971, not long after graduating from the State University of New York at Buffalo with an engineering degree, he began science reporting for NPR and has been there since.

Now 51, he lives in Connecticut with his wife, Miriam, and their three children.

— *How do you describe what you do?*

I'm a yenta. I can't wait to learn new things. And then to tell people about them.

— *As a science yenta, how do you make complex subjects simple enough for your listeners to understand?*

First, I have to understand the story. And once I understand, I can act like a transducer and translate it in plain English.

— *When you're broadcasting, you break into your guests' statements frequently. Didn't your mother teach you manners?*

She didn't tell me not to interrupt. (Laughs) Look, I'm trying to put science into English. I have a rule for guests: "Stay away from jargon as much as you can because I'll be breaking in and interrupting."

I tell them: "I want you to make believe we are in your kitchen sitting around having a nice chat." Then, I come into the studio with two dozen questions in front of me. Most of the time, we never get to them. A successful show takes on a life of its own. My goal is for people to see that science can be as interesting a conversational topic as Jennifer Aniston's wedding or stock prices.

___ *As a child, did you grow up dreaming of someday becoming Mr. Wizard, the television personality of the 1950s who introduced baby boomers to the joys of science?*

I practiced it. I was the kind of kid who did experiments in basements. I'd travel to Vesey Street in Manhattan and go through electronics stores to pick up old radar parts and transistors. Then back at home, I'd take a heavy-duty wire coil, plug the wire into a wall socket and turn it until it glowed. That basically turned it into a toaster. I'd get transformers and hook them up backward and get mushroom clouds out of them by plugging them into the walls and then melting them.

I liked plugging stuff into the walls. As a kid, I almost blew up my mother's bathroom trying to take chlorophyll out of leaves. I was more worried about my mother finding out than I was about the fact that I'd almost burned down the house.

___ *How do your listeners differ from those of, let's say, Rush Limbaugh?*

Our audience runs the gamut: parking attendants, science geeks, professors. When we got our first call from a homeless person, I was impressed. He asked a very good, very intellectual question. More often, we get professors because they want to hear what colleagues are up to.

On the other hand, I sometimes get callers who are scientists with contrary ideas to those of my guests. They'll call in and complain, "I've been trying to reach you for years, and you won't return my call. Why won't you read my paper?"

___ *How do you make science visual on the radio?*

You create shtick. I once got a letter from a woman in Baltimore who wrote, "My kids go into closets and they chew wintergreen Lifesavers because they spark in the dark. What's this about?" So Susan Stamberg and I took a microphone into a closet at the studio and starting chewing wintergreen Lifesavers while describing to the listeners what was happening. "If you folks out there understand this phenomenon, call us and tell us about it," we said.

The phone rang off the hook. "It's something called tribo luminescence. When you chew on a sugar crystal, the crystal breaks down and creates a little spark," one listener explained. "Francis Bacon wrote about this hundreds of years ago because he used to keep sugar in a giant lump in the basement," another told us.

Another time, a tanker ran aground off Nantucket, and they were talking about whether or not all this oil was going to foul up the ocean. So I said, "Let's create our own oil spill." I got a 10-gallon fish tank, sand, artificial fish, and some No. 6 heating oil. We created this oil spill in our tank and talked about it: "Oh look, there are globules and some of them are sinking down to the sand! They are getting stuck on the sand just like they say is happening in New England."

— *Why is there so little science on commercial television?*

There's a lot of medicine on the air, but not much science. For science to get on commercial television, it has to be sensational, polarized.

I think part of the problem in TV is the media gatekeepers: they're the people in charge of putting people like me on the air. And they are often people, I believe, who had trouble with science when they were in school. They just don't think science is terribly interesting.

— *Who's your favorite interview subject?*

Leon Lederman, the physicist. He's got a Nobel, so he's got credentials. He's not afraid to speak in a language we all understand. Even though he's a serious scientist, he doesn't take himself seriously. Most important, he's self-confident enough that he's willing to admit that he doesn't know everything.

— *What made Carl Sagan so terrific a science communicator?*

He was willing to suffer the slings and arrows of his colleagues. Carl Sagan was despised by many in his field because he was a popularizer.

— *How critical of science policy are you willing to be on your show?*

We're critical that scientists are not proactive, that when they really believe in something they don't take a stand. We're critical about where funding goes.

— *One of the show's funders is the National Science Foundation. Does the N.S.F.'s generosity create any conflict when you cover their work?*

If there is a conflict, we always mention that the N.S.F. backs the show. But that tie is going to end this year. At the end of this year, we'll be totally backed by NPR. Actually, there isn't so much a problem with them as some of the other funders of NPR. NPR has underwriters who are from biotechnology companies and we are doing more and more stories on biotechnology. Sometimes we'll have guests who are funders. When that happens, we say it up front.

— *Do you consider yourself as a kind of missionary for science?*

No. Because missionaries have to convert people. I don't have to convert anybody. People love science to begin with, and they'll take as much of it as you'll give them.
April 4, 2000

Postscript

Journalists and writers are often the best interview subjects. They know what makes a story work—and in their collegial way, they try to help out. Thus, interviewing science journalist Ira Flatow wasn't work at all. He simply turned up at my apartment in Chelsea in New York City, accepted a cup of coffee, and began sweetly schmoozing. In no time, we had an interview that was good to go.

A year later, Ira Flatow reports via e-mail that of course, he's still doing *Science Friday* on National Public Radio, but he is also "voicing three one-hour specials on the Discovery Channel. I am guest lecturing at the Columbia University Journalism School and moderating a journalism panel at Harvard, a school without a Journalism School. On the Web, we continue expanding and enlarging our sciencefriday.com Web site, hoping if we can get the funding to turn it into a multimedia science portal. We had been producing weekly teaching materials, curricula from *Science Friday* programs." He also noted in a second e-mail that "the bad news is that since the last time we chatted, I haven't won the lottery, but the good news is that I won two major awards— the 2000 American Association for the Advancement of Science Journalism Award and the Brady Washburn Award from the Boston Museum of Science."

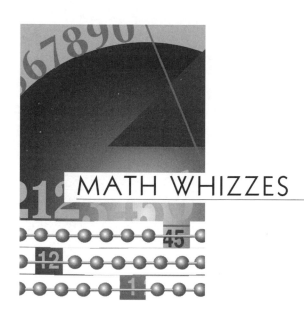

MATH WHIZZES

SIR ROGER PENROSE

A Mathematician at Play in the Fields of Space-Time

Jonathan Player

The impish man sitting in a warren in a remote corner of the mathematics building at Oxford University in England is **Sir Roger Penrose,** 68, mathematician, physicist, author, teacher—perhaps one of the greatest living disciples of Albert Einstein. In the 1960s, Dr. Penrose laid much of the foundation for the modern theory of black holes—objects so dense, according to Einstein's theory of general relativity, that they would collapse space around them, creating a hole from which not even light could escape. He is also the inventor of so-called Penrose Tiles, which can cover an infinite plane without ever repeating their pattern.

Americans know Dr. Penrose from his books on scientific questions, including *The Emperor's New Mind* (Oxford University Press, 1989), which criticizes the idea of artificial intelligence, and *The Nature of Space and Time* (Princeton University Press, 1986), which he wrote with his friend Stephen Hawking.

___ *In the last year, there has been an accelerating pace of discoveries in physics and cosmology. Which have impressed you most?*

Some of the most impressive things, I think, are pictures the Hubble telescope has sent down where the gravitational lens effects have been observed. This is the bending of light through Einstein's theory of general relativity, which tells you that light is bent by a massive body such as the Sun. The Sun bends light and acts as a lens. You can detect distributions of mass this way. You may not be able actually to see the mass in this big galaxy, but if there are other things behind it, you can see its lensing effect.

___ *Were you surprised by some of the recent discoveries confirming the existence of black holes?*

"Surprised" would be a little too strong. I think recent evidence is pretty potent that black holes are there. For a long time, the evidence was rather indirect. But you knew there was an object out there that was too massive to be anything else.

143

— *So how are you voting on the shape of the universe?*

Well, there are three main models. One, is positive spatial curvature—that the universe is like a huge sphere. Two, is zero spatial curvature—a so-called flat universe. Three is negative, "saddle-shaped" spatial curvature. I go with saddle shaped. For a host of reasons, I think that's the most likely. And the evidence is in fact shifting toward that. The universe seems "low density" and therefore of negative spatial curvature.

In terms of new discoveries, what I wouldn't like is the cosmological constant—the mysterious cosmic repulsion Einstein inserted into his equations to keep the universe stable—being nonzero. I've worked in general relativity for most of my life, and there are aspects to the theory, which when you don't have the cosmological constant, are much more appealing—at least to me.

Now, if they find that, yes, there has to be a cosmological constant, I'd have to go along with that. I'm not saying that I've closed my mind to it. I would just prefer the cosmological constant not to be there.

— *And if the cosmological constant is there—what does that do to your work?*

Complicate it! (Laughs)

— *You seem to have a good capacity to live with being wrong.*

Well, a scientist should have. If I'm really shamefully wrong, I have to adjust.

— *When you're doing some way-out mathematical computations, do you sometimes feel like an explorer?*

Yes, but also like an archeologist. You are looking for clues that are sort of lying there, hidden. And you see different forms of what may be underneath there, and you see some of the corners, and you have a sense that there may be some structure that may have influence on various different areas. Then you dig away until you can find what the structure is.

— *There was a debate in this newspaper last fall, "the end of science." Do you agree with those who argue that when it comes to physics and mathematics we know just about all that can be known?*

The argument is that accessible important ideas will run out, that either we'll never understand certain things or we have already understood them. This is an absurd position to take. People have been saying that for centuries.

I can think of a least one major area, which I'm absolutely sure is missing from the present-day physics, which probably will come in the next 50 years or so, and it will be a tremendous revolution. It has to do with how to understand quantum mechanics. See, quantum mechanics describes small-scale phenomena—atoms, molecules, particles. And you have certain rules, which if you try to apply them to large objects, they give you nonsense. They will tell you that a baseball can be here and there at the same time.

There are endless ways that people try to argue around this. But to me, it says that the theory is just not right and that there is a level that things go over from the quantum to the classical. I think one can make good estimates as to what level that is. And I even have proposals for experiments, which I hope will be performed, which would tell whether this is right or not.

— *Among many physicists today, the leading candidate for Einstein's dream of a unified "theory of everything" is the theory of superstrings, based on the notion that elementary particles are not points but strings?*

String theory is an example of science being driven by fashion. And I have mixed feelings about it. Some of the mathematical notions which people associate with string theory are very appealing. But just because they are appealing doesn't mean that they are right. In a book I'm trying to write now, I want to talk about all these theories, including my own baby, twistor theory, a mathematical scheme for re-expressing the structure of space-time in which individual points are not regarded as the fundamental things, but entire light rays are taken as more fundamental as individual points. Sorry. I can't explain it in a less complicated way. My point in the book will be that twistor theory certainly hasn't solved some of the problems that I thought it would, but it's very much alive.

— *Ten years ago, in* The Emperor's New Mind, *you said that computers were not likely to ever have consciousness—that all they could do was merely "compute." With all the recent advances in computer technology, do you still feel that way?*

Yes. A computer is a great device because it enables you to do anything which is automatic, anything that you don't need your understanding for. Understanding is outside a computer. It doesn't understand. Whereas to know what the calculations are supposed to do, what the answers mean when it's finished, requires your understanding and that's complementary to the computation.

And the understanding aspect of it is something that requires one's awareness, consciousness. If you didn't have consciousness, you couldn't understand. The strong artificial intelligence people feel that consciousness is some kind of "emergent" phenomenon; it's something that just comes about.

_ *Why do you think some of the artificial intelligence proponents seem to have such a stake in the notion that a machine can outthink a human being?*

I think that one reason is that people really cannot think of anything else which they would call "scientific." And these computing devices are tremendous, and they can do computations in a fantastic way. But the artificial intelligence proponents think that's all there is. They think that a scientific theory has to be computational. If you come from mathematics, as I do, you realize that there are many problems, even classical problems, which cannot be solved by computation alone. Another reason is people get work, they get grants, they publish papers, and they want to get more money. So they don't like my coming along and saying, "No, no, there are limits to what you can achieve this way."

The third is, it's a bit like religion. The question of what the mind, what consciousness is, is related to what people want religion for. Some religious people feel happy with a certain viewpoint, and they don't like to be challenged. These artificial intelligence people are somehow happy because they have a model that they are content with, and they don't want somebody coming around that disturbs that.

_ *In the match a while ago between IBM's "Deep Blue" computer and Garry Kasparov, for whom were you rooting?*

I was rooting for Kasparov, human chauvinist that I am. I'm not sure that he was the right person to play the match because he depends too much on being brilliant. I think (Anatoly) Karpov, who plays more consistently, but not so brilliantly, would be a much better person.

— *So you don't think that anyone will ever be able create computers with emotions?*

No. I believe there is something going on in a conscious being, which includes many animals, as well as ourselves, that is not a computational activity. And to be conscious at all is not a quality that a computer as such will ever possess—no matter how complicated, no matter how well it plays chess or any of these things. It doesn't mean, on the other hand, that somebody sometime in the future could not build some kind of a device, NOT a computer, which did whatever we do. I'm saying we won't get at it through computation alone.

— *What are you working on now?*

Two major research areas. The first is twistor theory, and more specifically, in developing its relation to Einstein's general relativity, which I think is key. Once that connection is made clearer, then the hope is it will give good leads on how to combine general relativity with quantum mechanics. There's a big gap there. If twistor theory can incorporate general relativity then it will give leads on how to unite general relativity with quantum theory.

The other thing I'm working on also has to do with the measurement problem in quantum mechanics. It's the link I mentioned earlier between the small scale and the large scale and the possibility of experiments that will show the limitations of present-day quantum mechanics. This is an experiment that I hope will actually be performed in space. It's much more directly related to observation than twistor theory, which is mathematical. If the experiment comes out the way I hope, it will tell us we need a major revolution in quantum mechanics.

— *You have toys here in your office. Why?*

(Laughs) Science and fun cannot be separated.
January 18, 1999

Postscript

On a gorgeous evening in November of 1999, I sat in an opulent room at the London Park Lane Hilton and wept. The next day, I was scheduled to interview one of the world's great mathematicians, Sir Roger Penrose, at his office in Oxford, and I was frightened out of my mind. This was the situation I had been afraid of when I accepted the interview gig at *Science Times*. Here I was thousands of miles from home and way over my head.

Long ago, in the 1960s, when I was a rebellious teenager growing up in Brooklyn, I angered my mathematically-inclined father by failing high school geometry four times. As a kid, I looked at mathematics as if it were an alien language from deep outer space. I picked my college, New York University, on the basis that at the time, it had no math requirement.

So there I was, an adult woman staring at Sir Roger's books and reliving the sorts of childhood failures that one usually brings into a psychoanalyst's office. Desperate, I ordered coffee from room service and began going through the Penrose oeuvre. "Treat it like a foreign language," I told myself. "Do what you do when you go to a foreign culture and don't know anything. Immerse yourself and see what you can instinctually grasp. Kid, just pretend you've landed in Burma!"

Amazingly, the strategy worked. Somehow, I struggled to the essence of many of Penrose's ideas. At Oxford, the next day, the celebrated mathematician I encountered was very kind, very open—sweet. Yes, sweet. That's the right word for him. It's a cliché to say that the best artists and scientists have their "inner child" at close hand, but with Roger Penrose it was really true. He worked in an office filled with toys which he kept, he said, because he enjoyed playing with them. For me, the toys broke the ice. I ceased being afraid of him, as well as the subject. The interview became improvisational, something like jazz. It was freeing in being in such unfamiliar territory and feeling my way around it. Penrose's kindness and accessibility made the experimentation possible.

This is one of my favorite pieces in a thirty-year career as an interviewer.

Two and a half years after our meeting, I wrote Penrose to see what was new in his life.

His response: "I suspect that the momentous event that has happened since then was the birth of my son, Max, on May 26, 2000, who is a delight. There were some other things too, receiving the OM (Order of Merit), having an art exhibition of my drawings, achieving some new results in a quantum-mechanical experiment, and still not having finished my new book.

Best wishes, Roger."

LEON M. LEDERMAN

Science Is Serious Business to the "Mel Brooks of Physics"

Steve Kagan

The small, loquacious man lighting into a slice of rhubarb pie at a South Loop restaurant on a recent Chicago evening is **Leon M. Lederman, 76,** City College class of 1943, director emeritus of the Fermilab National Accelerator Laboratory, author of *The God Particle* (Houghton Mifflin, 1993) and winner of the 1988 Nobel Prize in Physics.

In the world of science, Dr. Lederman is known for his detection of the muon neutrino, the subatomic particle that is one of the essential building blocks of matter and for the fact that he may be the wittiest particle physicist extant.

Since his retirement from Fermilab in Illinois, Dr. Lederman has become an advocate for improved science education. He is chairman of the board of the Teachers Academy for Mathematics and Science, an organization that shows primary school teachers how to teach modern mathematics and science.

— *You've been called "the Mel Brooks of the physics world." Brooks's humor is rooted in fear—fear of poverty, fear of the Holocaust. Where does yours come from?*

It comes from a terror of taking myself seriously. . . . It also puts me into a comfortable relationship with other people. It's human to want to make people laugh. It's part of teaching. Teaching is show business. Once, I gave a talk before the New York Science Teachers Association, and we met at the Stardust Room at the Nevele Hotel (in the Catskills). And I knew all about that place from my childhood, from listening to the great comics like Zero Mostel who broadcast from there. And I was so thrilled to be there that I actually took the microphone off the stand and started telling jokes: "The teacher says to the student, 'Hey, wake up that kid sleeping next to you,' and the student says, 'Why should I? You put him to sleep!'"

149

— *You worry about students falling asleep in science classrooms. In fact, you are a major advocate of reforming science education. What have you learned from your efforts to change the way science is taught in our public schools?*

That resistance to change is awesome. I believe we're teaching high school science in the wrong order—biology, chemistry, and then, for 20 percent of the students, eventually physics. In this sequence, the subjects are unrelated, to be learned and forgotten in the order taken. It would be much smarter to teach physics in the ninth grade, which would teach atomic structure to provide an understanding of how atoms combine, the essence of chemistry.

Modern biology, which should be taught last, is molecular based. I once suggested this to a group of high school physics teachers and thought they'd applaud wildly. They were ice cold. Then, it came out: "You want us to teach freshmen? We don't do freshmen—we do seniors!" Tells you a lot, doesn't it?

— *In Einstein's time, physicists were celebrities. Do you wish that were true today?*

Not really—though I sometimes tire of being introduced as the "1988 Nobel Prize winner for physics," and though I do believe science needs a better press.

Toward that end, I've been trying to sell a TV series, which we called *L.A. Science*. We'd have the usual sex and drama and car chases, but the hero would be a scientist, and every segment would teach some real thing about science.

Lawrence Tisch, the former owner of CBS, advised us on how to draft the proposal—to write up something that those guys call a bible, which includes sample story ideas, characters, etc. We saw Steve Bochco (producer of *N.Y.P.D. Blue*). His reaction was more or less, "Gee, this is really good, but I don't do scientists."

We tried Chris Carter *(The X-Files)*, who said something like, "I have my own ideas." In fact, he didn't want to be burdened by having every scientific point be plausible. All things being equal, he went for the magic and the surreal and what we call junk science. We tried Michael Crichton *(ER)*, who quickly dismissed us because he's anti-science. But we knew that from seeing *Jurassic Park*. We're still trying.

— *You won the Nobel Prize in Physics for your discovery of the muon neutrino. What was your reaction to the news last month that a group of Japanese scientists have shown that neutrinos have mass?*

I was pleased because it was really an elegant piece of work. The significance depends on knowing more than we now know about the numbers. What's nice is that this result energizes the many neutrino experiments under construction.

Since the neutrino has zero charge, zero radius, zero strong coupling . . . having a mass raises it to the level of significance. The new question now appears: What is the origin of this mass? In the spooky world of the quantum, the neutrino mass may be an indication of "presences" which are not in our inventory. . . . Just as the presence of a magnet seems to alter the properties of a nail, the existence of an unknown field could influence the neutrino's mass.

— *When you're investigating the micro-mysteries of the universe, how much does thinking about God and creation come into play with your work? I ask because you titled your last book,* The God Particle.

It comes into play. It doesn't have to. But it does. People ask you about it. There's always a place at the edge of our knowledge, where what's beyond is unimaginable, and that edge, of course, moves. . . . We've gotten closer and closer to some ultimate question. We now know that the universe was created some 13 billion years ago, and we will get that number more precisely, in some kind of fiery explosion called the Big Bang. The question is: Is that creation?

And then, "What was there before?" A possibly defensible answer is: the laws of physics. They dictated there should be this event. Then, if you want to go in that direction, some will ask, "Well, how did the laws of physics get there?" And then, you're stuck. I usually say, "Go across the street to the theology school, and ask those guys, because I don't know."

What I believe is that the laws of physics got there, I don't know how, and they determine the future course of the universe. The sequence is: the Big Bang, expansion and cooling, the formation of complex objects—eventually atoms, and the atoms formed molecules, and the molecules formed things that crawled out of the ocean. And here we are, worrying about the whole thing!

_____ *In the 1980s, you were the prime mover behind getting the United States Government to start building the world's largest superconducting supercollider in Texas. This would have been used for the most advanced of particle physics investigations. Then in 1993, after spending $2 billion on the project, Congress canceled it. What happened?*

It was politics and policy. . . . The scientific community underestimated, I think, the difficulty of starting a brand new laboratory. . . . The lab and its supporters were not politically astute enough. Defense industries, desperate for new sources of business, swarmed on this young laboratory, with their own ethic which is very different from that which had marked scientific constructions before. In industry, they work with underbidding and finally getting their money back with supplemental requests for funds and (that ethic resulted in a conflict between them and the scientist-managers at the lab). And then, deficit reduction became a large Congressional concern. Congress knew there were conflicts, and the whole project was expensive. The lab became a juicy target. I still think we should have built this machine. It was designed to do a certain job, and it was 20 percent complete.

_____ *What became of the $2 billion invested in the supercollider?*

It went into a hole in the ground. There's a big hole in Texas, probably filling with water. We probably recovered about a half a billion dollars worth of stuff that could be used elsewhere and unknown value for the information that has been published. We learned a lot. But most of it was lost. . . . One of the things we learned about this is that when you're trying to probe the subtle ingredients that make the universe, it's expensive. You'd better start building these things internationally, from the beginning.

_____ *Do you miss the cold war, at least as a funding wedge?*

(Laughs) I guess. It's a good question because here we always thought, naively, that here we are working in abstract, absolutely useless research and once the cold war ended, we wouldn't have to fight for resources. Instead, we found we were the cold war. We'd been getting all this money for quark research because our leaders decided that science, even useless science, was a component of the cold war. As soon as it was over, they didn't need science. We're fighting that presumption now. It's a tough job. You have to educate. That's one reason we wanted a television show.

July 14, 1998

Postscript

What has Leon Lederman been up to two-and-a-half years after our interview? The same old stuff but more so. In an update on his life sent on February 7, 2001, Lederman seemed hard at work at his mission of reforming science and math education in the United States:

"I have been 'living' at the Illinois Mathematics and Science Academic, the public residential high school for gifted students I founded in Aurora, Illinois. . . . In parallel with this is an obsession with science curricula in high schools: the absurdity of ninth-grade biology followed by chemistry and then physics. That 99 percent of U.S. high schools still follow this 100-year-old sequence is symptomatic of the resistance of schools to change. We have organized ARISE (American Renaissance in Science Education) to see if we can bring rationality to science education.

"Finally, to come to grips with the future of High Energy Physics in the U.S., we, at the Illinois Institute of Technology, have established a consortium of universities devoted to studying advanced concepts in the design of future particle accelerators. This is unusual in that the standard university reaction is: Show me where the beam comes out." Dr. Lederman also has a new book to be published next year by Dell—tentatively titled *Quantum Science for Poets, Politicians and Policemen.*

CHARLES BRENNER

A Math Sleuth Whose Secret Weapon Is Statistics

Peter DaSilva

Oakland, California

In a time when full-time academic appointments are hard to come by, some mathematics Ph.D.'s may consider the career path taken by **Charles Brenner,** 55, one of the world's first fully employed freelance mathematicians.

Dr. Brenner calls himself a forensic mathematician, and his business card reads, "Charles H. Brenner, Ph.D., Aphorisms, Inferences, and Conclusions from Thin Air."

As part of his work with his consulting company, DNA-View, Dr. Brenner travels around the globe helping scientists do the mathematical calculations necessary to analyze DNA evidence. He also advises in paternity cases.

Dr. Brenner has always been a math rebel. In the 1970s, a time when he should have been finishing his graduate studies, he went to London where he supported himself for six years by playing bridge.

He returned to the United States in 1974 to get his doctorate in number theory from the University of California at Los Angeles.

After completing his degree in 1984, he found that some software consulting he had been doing over the years seemed to evolve into new work in mathematical consulting associated with DNA investigations.

"I like teaching as an academic visitor," Dr. Brenner explained on a recent afternoon while sitting in his office in his home here.

"But in many places, the academic environment can be stultifying," he said. "So this is a wonderful alternative. I get to travel the world, encounter different cultures, rub shoulders with some great scientists, and I get to look at vexing human problems."

— *What exactly is a forensic mathematician?*

It's a term that I invented. It has to do with the application of mathematics in the courts. Virtually all I do is DNA identification. The most obvious example is I work out probabilities.

155

If you have, for instance, genetic material derived from the crime scene, you have a suspect, and the DNA matches, the question is "Do they match by coincidence, and if so, how big a coincidence is it?" There are population studies that show how common various DNA traits are. In the case of stain matches, all I have to do is multiply together a bunch of probabilities in order to get an answer.

I've worked for defense lawyers and for prosecutors. Also, a lot of what I do is in paternity cases.

__ *Your consultations for a Filipino court in the paternity-inheritance case of the late California millionaire Larry Hillblom made several Amerasian orphans very happy and rich. How did that come to happen?*

A judge in Saipan asked me to advise. Hillblom had a fortune. He'd used it to indulge his taste for sex with young Asian women. After Hillblom died, there were all these young children scattered all over the Asian-Pacific world. They were potential heirs because, according to California law—he was a Californian—they had not been specifically disinherited in a will.

So to prove paternity, samples of Larry Hillblom's DNA were needed. However, when an investigator went to Hillblom's Asian homes, strangely no genetic material could be found. No hair, no toothbrushes, none of the normal things that are usual in any home.

Blood samples of these children and their mothers were then assembled. None from Hillblom's California relatives were available for that. The relatives wouldn't cough up any blood.

There was a hospital in California that had a mole of Hillblom's, but they refused to produce it. They subsequently produced the wrong mole. It was all very strange.

Given it was the only possibility, I designed a computer program to look at the genetics of these children. Of the DNA profiles that came to me, most of the children proved to have a similarity that could only be explained if they had the same father. That, along with evidence of a relationship between the mothers and Hillblom, won the orphans a lot of money.

__ *Software you designed has also been used in reuniting families separated by the 1970s "dirty war" in Argentina. How does that work?*

Well, you know, in Argentina, there are about 300 adults who were infants in the 1970s and whose parents were murdered by the military. These orphans were subsequently given to childless military families to adopt. I've designed software, used by the laboratory there, which helps make genetic identifications by looking at the genes of still-living grandparents.

A couple of weeks ago, I got the case of an eminent Uruguayan, a Nobel laureate, named Juan Gelman. His son and daughter-in-law fled to Argentina from Uruguay in the 1970s, and they were killed there as "terrorists." Their little child was apparently auctioned off to someone in the military.

Through underground sources, Gelman was eventually able to find out what happened to his grandchild. He was pretty sure a certain young Argentine girl from a military family was her.

The girl already had her own suspicions because her adoptive father, on his deathbed, said, "I hope you can forgive me," though he didn't say more. Gelman contacted the girl and gradually revealed what he believed to be their kinship.

My job was to make the calculation that nailed it down. I did the calculations, and the data was overwhelmingly consistent that this girl was his grandchild.

— *The immigration services of several different countries have bought your software. Why?*

Well, in Denmark for instance, there is a law where if one person has gained legal entry, then certain relatives are allowed to immigrate, too. Well, in Somalia, where a lot of immigrants seem to come from, there's a very broad definition of "family." A potential Somali immigrant to Denmark might feel he or she is entitled because of membership in the same clan. When that happens, the Danish authorities go to a DNA laboratory that uses my software and who I consult for, to check the odds of a direct relationship.

— *What is the most interesting thing you've learned about human beings from your work?*

That quite a bit of adultery is intrafamily. One of the problems that paternity labs come up with is where a man is accused of paternity, and upon testing, it seems that he shares a lot of genetic similarity with the child in question, but not enough to be that child's father. What is going on here?

In those cases, you always wonder if the man being sued is not a father, might he not be the uncle! An uncle would explain this evidence very well. So what we have is a woman who is having relationships with a husband and the husband's brother. My colleague Jeff Morris and I did an analysis some years ago in which we came to the conclusion that this kind of relationship happens in at least 1 percent to 10 percent of cases.

— *How did you develop this odd but interesting profession?*

I come from a mathematical family. My father taught, and my mother was an artist and a politician—she was mayor of Palo Alto. In my family, practical mathematics was frowned on—statisticians, we didn't even talk of them. I find it rather amusing that I now do very practical mathematics.

But my history is this: I was in England during the Vietnam War, avoiding the draft and making a kind of living by playing bridge for money. By the time the war was over, I returned to the U.S. and graduate school, and I started developing a career for myself as a software writer. The software writing led to creating this DNA software. It's become a full-time business—the software, consulting, teaching. Basically, I'm a freelance mathematician.

— *Is freelance mathematics lucrative?*

It's not bad. I can do what I want, but I make less than most senior academics.

— *The murder case where most of us learned about DNA evidence was that of O. J. Simpson. How would you rate the presentation of DNA issues during that trial?*

The defense did something very clever from the DNA point of view: They said the evidence was planted. Their basic strategy was even if it matches, it was a plant. They gave up on the strategy of disproving the DNA evidence. There obviously was a match in the blood. They never denied it.

— *What did you think when you first heard of Monica Lewinsky's unlaundered blue dress?*

I tried not to pay attention. I was impressed by Bill Clinton's toughness in the situation, but it did strike me that if they could get a typing, his goose was cooked.

— *Does your work give you satisfaction?*

Sure. We are always reading about people who are very critical about the justice system. With this kind of work, at least you have something you can hold on to. It's a tool that can make the system more just.
August 8, 2000

Postscript

I first heard about Charles Brenner in the spring of 2000, when editor Dennis Overbye suggested that I check out his Web site, DNA_view.com. A few clicks later, I was in the world of this playful Sherlock Holmes and I could see in an instant, he'd be a marvelous interview subject.

On Brenner's Web site were his musings about several sensational murder and paternity cases, an enigmatic "quote for the day," some personal thoughts on Thomas Jefferson's sexual life, and some anecdotes about the venerable card game contract bridge. This was one quirky guy.

Brenner's originality apparently came right through the page when my interview with him finally appeared in an August 2000 *New York Times*. Suddenly, an agent from International Creative Management was calling me, not about my work, but for the mathematician's phone number.

In February of 2001, a year and a half after the interview was printed, Brenner wrote me that, "nothing terrifically new and different is happening with my work. We, my girlfriend Nancy and I, recently returned from Malaysia where I helped train the DNA lab in the use of my software and gave some lectures in the associated mathematical theory. The stopover in Vietnam was new and different, but not professional.

"I have an ambition to set—or at worst, tie—a record by working for both sides on a local ice pick murder. Murder case, rather. The murder is over and done with for thirteen years so I have no possibility to participate. The prosecution has engaged me to bolster the DNA evidence, which is good evidence.

The defense has raised objections to the fingerprint evidence, and I think this could be a kind of application—a cold hit in a database search—where an enterprising forensic mathematician could make a sound argument that there are untested and unjustified underlying assumptions. The defense attorney and I are old nemeses."

Dr. Brenner also reports that nothing ever came of his brief dealings with the International Creative Management agency.

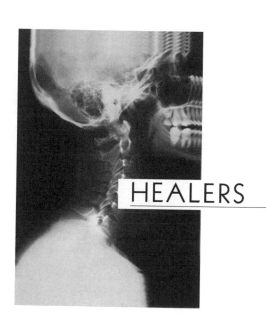

HEALERS

NANCY PADIAN

Battling AIDS in Africa by Empowering Women

Peter DaSilva

San Francisco

Lounging in a high-floor room at the St. Francis Hotel, drinking coffee served in perfect porcelain cups, **Nancy Padian** was far from the sad, overcrowded streets of Harare, Zimbabwe, but they and their beleaguered people remained on her mind.

Dr. Padian, 48, an epidemiologist who is director of research for the AIDS Research Institute of the University of California at San Francisco, is one the world's foremost experts on the heterosexual transmission of AIDS.

For the past seven years, she has commuted between San Francisco and Harare—some 10,000 miles—to help Zimbabwean health workers gain some measure of control over the AIDS epidemic tearing through their nation in southern Africa. Their chief weapon against HIV has been the barrier method of contraception, using devices like condoms.

"There's this zeal to find this new magic bullet," Dr. Padian said. "But there are all these existing methods with great potential, and they are right under our noses and that haven't really been looked at. I'm hoping that what we do in Harare will help the women there and, frankly, here too."

— *You've racked up thousands of frequent flier miles between San Francisco and Zimbabwe. What exactly do you do there?*

I run a series of studies as part of a joint project between the University of Zimbabwe and the University of California, San Francisco. We examine the effect of contraception in preventing HIV in Zimbabwean women. In Africa, AIDS is mainly heterosexually transmitted. So what we are looking at in a country where 25 to 30 percent of the population is infected with HIV is whether different forms of contraception might increase or decrease the acquisition of HIV.

We see a population of women who come to regular family planning clinics in Harare. About 30 to 40 percent are infected. The rate is much higher, of course, for women in the sex industry. In the one study, we try to help uninfected women to get their male partners to use condoms. We've been able to persuade people to use condoms far more than we originally thought we could. In fact, we've been able to get more than half the women in this category to do this.

For those women who aren't successful at that, we try to encourage them to use female methods of contraception: female condoms, spermicides, diaphragms. We are also looking at whether or not women who are using hormonal contraceptives are at increased or decreased risk of HIV. This study has just begun, so we don't have any answers yet.

— *The assumption has been that women in Africa and elsewhere cannot convince men to use condoms. What are you doing differently?*

Some women can't. But for others, it's a question of teaching them negotiating strategies. We do role-playing. We present the women with obstacles and have them work out ways to overcome them. We encourage them to think it's O.K. to talk about sexual activity with their partners, that these discussions are part of a healthy lifestyle.

— *What makes women in southern Africa so much more vulnerable to AIDS than American women?*

There is a higher prevalence of other sexually transmitted diseases there, as well as a lack of male circumcision, which turns out to be a factor in the susceptibility of the man to AIDS. And you know, the rates of infection from parasites, poor nutrition, all contribute to the problem. In general, if a woman's immune system is compromised because of other infections and nutritional factors, that too increases susceptibility to HIV once she's exposed.

There are some particular cultural factors, too. It is not uncommon also for African men to have multiple partners and still have their monogamous partner. The practice of "dry sex," which occurs in certain countries in southern Africa, almost certainly increases the susceptibility of women to HIV, if their partner is infected. Dry sex is when a woman uses a chemical or physical substance that she puts into herself to dry up the cervical and vaginal secretions in advance of having sex. Women sometimes engage in this practice because there's a belief that men prefer it.

— The Colombian writer Silvana Paternostro claimed that the biggest AIDS risk factor for a woman in many Latin American countries was to be married. Is that true in southern Africa, too?

I think that's true in Zimbabwe, where the infection rate is from 25 to 30 percent, and where the likelihood that a husband is infected is quite high. It's not a question of the woman having multiple partners. Unless they are sex workers, the women are generally monogamous.

— To what extent do you think the power imbalance between African men and women contributes to the spread of AIDS?

I think the gender imbalance is huge there and accounts for some of the transmission. One of the things I see as helping in preventing AIDS is microloan programs for women's economic development. These are small loans to women to help them start businesses of their own. In order for women to really negotiate sexual activity, it's important for them to be economically empowered.

— The president of South Africa, Thabo Mbeki, has issued some rather ambiguous statements on AIDS. He has raised doubts about whether HIV causes AIDS, declared a need for an African solution to the epidemic, and suggested that researchers should look at poverty as a root cause. Considering that the South African AIDS rate has increased 50 percent since his election, what do you believe he's thinking?

I don't understand why he said it. I wondered whether he was confusing two issues. I completely agree that we need indigenous solutions to indigenous problems. You definitely can't export American models of behavior changes and expect to implement them in southern Africa. But to suggest that HIV doesn't cause AIDS when all the evidence points to the contrary verges on being irresponsible. It delegitimizes prevention programs. Most prevention programs, after all, are geared at preventing someone from being exposed to HIV And if HIV is not the issue, then what are we doing?

— Is President Robert Mugabe of Zimbabwe in similar denial?

Denial may be too strong a word here. Mugabe has certainly not dealt strongly with the epidemic. He's not strongly supporting the national

AIDS control program in promoting condoms. Whether that's denial or avoidance, it's a shame because your best chance of success is if you can engage the local leaders. If the government isn't behind it, it's much more difficult to have widespread effective programs throughout the country.

___ *How do you deal, emotionally, with the high death rate you are witnessing in Harare?*

I find it unbelievable, overwhelming. Among my colleagues at the university, someone I am working with is always off to a funeral. It's unbelievable. But I feel like I'm doing something. And that keeps me going. What makes me angry is when people say that it's all sealed and that it's too late and that you can't do anything more. Because I feel like even if I do a little bit, it will make a difference.

___ *How did the epidemiology of HIV in women become your specialty?*

I was in graduate school here at Berkeley in the early 1980s. I had originally wanted to be a psychologist, but fell in love with epidemiology because it seemed to be a good way to help a lot of people—rather than one person at a time.

My graduate adviser, Warren Winkelstein, had a grant to look at the natural history of AIDS. HIV hadn't been discovered as the cause yet. So at that point, no graduate student was interested in working on this. I got the job as his assistant because there were no other takers. He suggested that I look into the female partners of bisexual infected men in the group he was studying. That's when I started getting involved in the heterosexual transmission of AIDS.

Of course, what we found was, at that time, that male-to-female sexual transmission was a lot more frequent than female to male, that risk factors for transmission were primarily not using condoms, anal intercourse, and STD's in either partner.

___ *You are about to go to India to start an AIDS study there. How bad is the problem on the subcontinent?*

The infection rate isn't that high, but the absolute number of people who are infected is huge. Huge! Given the numbers, one can only guess that the epidemic exists at an early stage. And, of course, all of the gender

issues that we talked about in Zimbabwe are acute there as well. The women have little say. What I'm discovering as I travel the world is that for women, power is an absolute dimension of physical health. The bottom line universally is, if you cannot negotiate what you are doing with your body, you will not be able to lead a healthy and long life.
August 22, 2000

Postscript

Nancy Padian sadly has no shortage of work to do.

"Since we spoke I have begun some feasibility/pilot work here in Zimbabwe, to lay the groundwork for adolescent girls to minimize the use of 'sugar daddies' or other forms of survival sex," she wrote me in February of 2001.

"Ultimately, the intervention will have three components: a piece to keep the girls in school (or get them there, if they are not) including provision of school and uniform fees and other material needs, some sort of job training, and then most likely a micro-financing package to launch them on their way to economic autonomy.

"This goes back to the issue we discussed about needing to be economically independent and empowered in order to be in charge of your own body and choices related to sexual activity and protection.

"Our current work focuses on identifying community partners, qualitative work to formalize the intervention and establishing our plan for the evaluation. I expect we'll be in the design phase of the intervention next fall and then our challenge will be to find funding."

NAWAL M. NOUR

A Life Devoted to Stopping the Suffering of Mutilation

Rick Friedman

Boston

In all the vast territory of the American health care system, there is nothing quite like the African Women's Health Practice of the obstetrics and gynecology department at Brigham and Women's Hospital here.

Every Friday, dozens of African women immigrants, now living in the area but still wearing the brilliant gowns of their native countries, come here for help with their medical problems. Many are seeking not just routine care in the specialty, but treatment for special gynecologic conditions related to female circumcision.

The ritual, which is also called female genital mutilation, is performed on young girls in many parts of Africa.

This unusual clinic is the brainchild of **Nawal M. Nour,** 34, a Sudanese-born, Harvard-trained gynecologist whose 1988 Brown University undergraduate thesis was "The Emancipation of the Egyptian Woman."

A crusader against female circumcision, Dr. Nour founded the clinic to help women who had been mutilated and to give her a platform to organize doctors and other professionals against the ritual.

On a recent cheery afternoon, when it seemed as if all of Boston was out along the Charles River enjoying the warm sun, Dr. Nour sat in her cubbyhole of an office at the Brigham and told how battling female circumcision had become her life's passion.

— *When did you first become aware of female circumcision?*

Early. I grew up in the Sudan, Egypt, and Great Britain, and so, for as long as I can remember, I was aware of it. My father is Sudanese. He was an agronomy professor and my mother is American. My parents were very influential. My father spoke out against it.

At school, I remember the girls saying: "I was circumcised. Have you been?" I remember one girl saying she'd been circumcised and that it hurt, but it was a good thing because now she was a woman.

The practice troubled me, but I was also intrigued by it because it's so horrible—and yet, it continues. As a child, I couldn't understand why people would do something that wasn't good for them. I think I became a physician so that I could find an effective way to attack it.

To this day, I get e-mails from schoolmates who write: "You were always talking about this practice. I remember in the 10th grade, you brought up this issue."

— *How did this African women's clinic begin?*

In 1995, after I began my residency here, I started attracting patients from the Sudan, Ethiopia, Somalia, and west Africa. I became known in the immigrant community around Boston as that "African woman doctor" at Brigham and Women's Hospital. Most of these women who came to me, obviously, had undergone female circumcision. So eventually, I went to people from the immigrant community and asked, "Would you like for me to open an African clinic for Africans?" People were very excited. We did a focus group study and asked where they wanted this clinic— West Roxbury, Mission Hill? People wanted to go to the Brigham. They felt that this was where the wealthy Americans go and they should go to the same hospital.

— *What kinds of clinical problems do your circumcised patients bring to you?*

The major complications are seen on women who have undergone Type 3 circumcision. Type 1 removes the clitoris—this is common in Ethiopia. Type 2 excises the clitoris and the inner vaginal lips, which may end up fusing together. Type 3 is removing the clitoris, the inner lips, the outer lips, then sewing everything together, leaving only a very small opening for urination and menses. This is mainly done among Somalis and Sudanese and in parts of west Africa. Female circumcision, you see, is nothing like what we know as male circumcision. In the latter, the foreskin is removed from the penis. With female circumcision, we have the equivalent of the penis being removed.

The women who've undergone Type 3 can have scarring problems and problems with their menses. Some of them have terrible trouble having sex with their husbands, as you can imagine.

It's still shocking to me. This morning, I saw an 18-year-old whose opening was about dime size. I saw a woman the other week who was pregnant and had a pencil-sized opening. One couldn't help but wonder how she'd managed to even get pregnant.

— *Can some of this be solved by plastic surgery?*

If the clitoris has been removed, it cannot be returned. But sometimes, when we do a procedure which opens up the scar tissue from the Type 3 circumcision the end result can look very good. It can look like new external lips have been formed. The women I've operated on are very pleased with that. They have less pain in intercourse and with their menses. It allows them to urinate quite easily, which often wasn't the case before.

If a patient is considering getting pregnant and she has a very small opening as a result of the circumcision, I encourage her to be opened up. If she is coming in with chronic urinary tract infections, pain on intercourse or pain in menses, I often suggest it, too.

— *Since you've had the clinic, have you gained any insight on a question you asked as a young woman: why does female circumcision persist?*

I think it persists for complex reasons. Some people perpetuate it because they say that religion requires it, though that's wrong. Islam never stated that girls should be circumcised. Others say they do it to make sure that their daughters will be marriageable. Still others argue that it's necessary to maintain female chastity. I find that people do it because of a deeply ingrained belief that they are protecting their daughters. This is not done to be hurtful, but out of love. The parents do it because they think this is necessary to ensure that their daughters will get married. They love their children. These are the same parents who in time of war or famine will give up their food so that the children will be fed.

One of the things our clinic tries to do is to educate American health care providers about the practice so that they can take better care of these women when they come across them. And then, I try to educate the women to prevent the practice from being perpetuated in the U.S. and to try to prevent them from sending their daughters home to be circumcised there.

— *In nineteenth century America, women were castrated and clitorectomized as a treatment for "hysteria" and "eroticism." These operations, some historians suggest, were often used as a means of social control. Is the same true in contemporary Africa?*

It's a very difficult question to answer. Some would argue that, of course, it is a means of social control over women. But the fact is that the people who are perpetuating the practice are usually the women themselves. That's why I think one of the things that must happen in Africa is that there should be dialogue between men and women. The men are not involved in it. It's "a woman's thing." There needs to be much more dialogue on how this affects a woman's life.

_ *How likely is it that an American gynecologist will ever see a circumcised woman?*

People in middle America may not see so much of it, but doctors in Portland, Oregon, Washington D.C., New York City, Seattle, Minnesota, and parts of California are likely to. Those are the places where immigrants are going from Africa. There are at least 12,000 Somalis in Boston.

_ *What do your patients tell you about their experiences with the American health care system?*

They often speak of going to a health care provider with abdominal pain and the provider does a complete physical exam and discovers that the woman is circumcised and is suddenly making comments like: "Oh, my God, what happened to you? This is the worst thing I've ever seen!" And then a patient in stirrups tries to explain it through an interpreter. All the while, she's thinking, "But what about my abdominal pain?" And the fact is, she may or may not have a problem with how her body looks, but she doesn't want to be judged. Given how she's treated, it's possible she'll never return to that provider again, even if she needs to.

_ *Ideally, how should medical professionals react when they encounter circumcised women?*

I understand that female circumcision is a horrible act and I empathize with the horror of the doctor, but what I ask is that a physician not reveal their emotions and thoughts to the patient. For people who haven't ever dealt with this, the whole thing may be difficult to understand. One can say, "It's a horrible practice and it needs to be stopped." But the practice is very different from the patient. The patient may or may not have wanted it herself, or she may be happy with the way her body looks. In any case, she should get sensitive and productive care.

— *What does your work with the circumcised African women do for you?*

There are days when I feel so fulfilled. When women come to me and tell me that their pain is gone, when they say "I didn't realize that sex didn't have to be painful," or they say, "I had so much fear about delivering my baby and it turned out to be so easy." Those are certain things I feel really great about.
July 11, 2000

Postscript

In the 1970s Women's Health Movement, a social movement that helped transform how medicine was practiced on both men and women, there often was much arguing over the question of what difference would more female physicians make to the overall status and health of American women? Hard as it is to believe in the year 2001, when 32.6 percent of all gynecologist-obstetricians are female, in the 1970s, the figure was more like 6.8 percent. And gynecology, at the time, was practiced in a way that frequently dismissed and infantilized women.

What the activists at that time wondered was: If large numbers of women entered medicine and gynecology would they merely advance their own economic interests, or would they become a professional army of advocates for their female patients?

School is still out on that question. But in the spring of 2000, I was leafing through a very useful newsletter that Harvard Medical School sends out to the press and found an article about Dr. Nour and her clinic for African women victims of genital mutilation. She looked like a great story.

Dr. Nour and I met at her offices at Brigham and Women's Hospital in Boston a few days later. In our two-hour interview, I got a sense of a woman using every opportunity the women's movement had made for her to improve the lives of women in need. She'd done everything right—Brown University, Columbia, Harvard Medical School—and the credentials she'd amassed were tools for helping others. She seemed to me very much the fulfillment of a thirty-year-old hope; in fact, it was hard to imagine a male doctor being quite as passionate and as effective in combating genital mutilation.

A year after the interview appeared, we spoke again on the phone. "The clinic has continued to flourish," Dr. Nour reported. "I've been operating more than usual. Word is out that I do this reconstructive surgery, and I'm doing more of it. We're changed the name of the clinic to the African Women's Health Center, which means that I will do more international work."

Toward that end, Dr. Nour said she'd recently given a presentation at the Faculty of Medicine of the University of Khartoum in the Sudan and "for many of them it the first time they faced the medical, social, and ethical issues surrounding female circumcision. I'm hoping to do some collaborative work with some of the NGOs (non-governmental organizations), actively working on it."

Mary A. Carter

JOHN BANCROFT

Sitting in the Ultimate Hot Seat:
The Kinsey Institute

For the last four years, **John Bancroft,** sexologist, sex therapist, psychiatrist, has been at the helm of the Kinsey Institute for Research in Sex, Gender, and Reproduction at Indiana University.

An Englishman with a grandfatherly manner and an Oxbridge accent, Dr. Bancroft came to Indiana from Western General Hospital in Edinburgh, after the institute became the focal point of a succession of political and management controversies. Dr. Bancroft's special assignment: rebuilding Kinsey's reputation and fighting off opponents who had criticized the institute.

Dr. Bancroft spoke recently in his Bloomington, Indiana, offices. Excerpts from a two-hour conversation follow.

— *I'm told one reason you were appointed at Kinsey was that you were good at dodging political minefields. Tell us about the topography of the minefields here at Indiana.*

It was, first of all, rebuilding the relationship between the Kinsey Institute and Indiana University, which got into difficulties over the previous few years. And getting the institute back on track as an active research institute. And getting credibility back, which had taken a bit of a beating because of controversies that had gone on.

There was a long and very painful, and for me, very mysterious conflict between my predecessor, June Reinisch, and the university. I don't fully understand it. I think June was treated badly, in a lot of respects. I think there was a campaign to discredit her. The university got, unwisely, caught up in this, and it was a very messy business, and I feel bad for June. But one consequence of that is that the Kinsey Institute lost a certain amount of credibility. Fortunately, the main players in that controversy here have moved on, and I'm able to deal with other people. So it was relatively easy to re-engage and get on with a relationship of trust and respect. The university has been great since I arrived.

— *Was part of Dr. Reinisch's problem that she was a woman dealing with that most unladylike of subjects, sex?*

She had a number of things against her. She was a woman, Jewish, from Brooklyn, working in a midwestern university, dealing with sex. June is a very colorful character, and she's very good at dealing with the media. She did what she was hired to do, I think, very well. But people started to think, "This is not what we wanted."

— *Among other projects, you are doing a study on how Viagra is affecting contemporary sexual behavior. Any tentative findings?*

We have just recently started a study. We're very interested in how the couple reacts. It will take us a good year. It's a very interesting drug. It's relatively free from problems. In general, I feel optimistic and positive about the Viagra phenomenon.

There are, of course, other aspects of it, one should feel more cautious about. It has fueled a huge movement in the pharmaceutical industry to find other ways of enhancing sexual behavior. The consequences of that commercially driven process are difficult to foresee. I have some anxieties about that. I'm a little anxious that it will serve to reinforce this phallocentricism of the male and his attitudes to male sexuality.

— *So you're saying that in pre-Viagra times, a sex therapist encountering an impotent man might try to get him more interested in foreplay and other nonphallic expressions of sexuality, but now, there are fewer incentives for that?*

That's right—show him that having an erection is only one part of lovemaking and that you can enjoy sexual intimacy in other ways. I'm not entirely pessimistic, though. One of the likely consequences is that people will realize once they have their erections back that there are other things they need to address in their sexual relationships.

— *What's your take on Bob Dole's Viagra ads?*

Well, I feel a lot of respect for him, actually. It was a courageous and very constructive thing to do. It certainly puts a new slant on the theme of politicians and sex.

— *Dr. Reinisch made the news in the midst of the Clinton impeachment when George Lundberg, the editor of* The Journal of the American Medical Association, *was fired because he published the results of a study she had prepared. In 1991, she asked 599 Indiana students about whether they considered oral sex to be sex and discovered that 59 percent of them, like Monica Lewinsky and President Clinton, did not.*

I thought that the firing was extraordinary. I think the actual data in the paper is a major contribution to knowledge. It threw some light on an issue that's very current. There was some advantage to having that evidence collected before this particular debate because it was therefore uninfluenced by it.

— *When you were a sex therapist in Great Britain, you did aversion therapy on homosexual patients. Do you regret that now?*

I don't regret it because I think my motives for doing it were entirely honorable. I just think it was a stage of development in the way we were thinking about it. I am embarrassed about it when people discover it.

— *Do you fear you damaged your patients?*

No, not at all.

— *Or perhaps, wasted their time?*

Oh, I think I must have wasted many patients' times over the years. Yes, I think it was a fruitless exercise, but it didn't take me very long to realize that.

— *Do you consider homosexuality a sickness?*

No. Not at all.

— *Then why do aversion therapy?*

Because one was responding to individuals' coming along and saying, "I want to change." I still respect that request. It's a request that one gets seldom or never, now. The individual who is wanting to deal with his sexual life, I'm there to try to help them sort it out, one way or the other.

— *It is said that you have one of the world's best collections of erotica here at Kinsey and that it is very difficult to get to see it.*

That's not true. There's a mythology around the institute. I mean, we have certain regulations we have to follow. That's one of the interesting paradoxes in the attacks against Kinsey: We are often attacked for not allowing people to come here. And yet, if we made the material that we have here generally available, there would be outrage, as well.

One of the things that made the collection relatively inaccessible was the extent that we didn't know what we had. And other things, like not knowing the agreements we had with donors and confidentiality agreements. It can be quite tricky. We have for example all the correspondence that Kinsey had. There are issues of confidentiality there. There are people's private collections, diaries, scrapbooks, photographs, fascinating stuff. We have an impressive collection of photography, art, films, and a pretty extensive library. But we are committed to making them available in the most appropriate way we can.

— *Were you picked for this job because you were the "staid" Englishman, a fatherly figure for Kinsey?*

That didn't do any harm. I think it needed someone who was reasonably mature. The interesting question is: "Who's going to be the next director?"
May 25, 1999

Postscript

Not long after I returned to my hotel from this interview, the telephone rang. It was Dr. John Bancroft. He'd been talking, he said, to his wife, and she pointed out how, *um,* he might have said some things that would, *err,* be hurtful to his predecessor at Kinsey, Dr. June Reinisch. He'd misspoken in the interview, he claimed. He'd been too relaxed, too unguarded. Would I consider cutting the entire section from the eventual piece?

I responded to this vaguely, with something like, "I'll think about it."

On the flight home, I listened to the portion of the interview that had left Dr. Bancroft so vexed. I had asked if his predecessor's problems at Indiana University had been partly rooted in the fact that she was a woman. His answer: "She had a number of things against her. She was a woman, Jewish, from Brooklyn, working in a midwestern university, dealing with sex."

Bingo. The problem here, it seemed to me, wasn't about Dr. Reinisch's feelings, the problem was in Bancroft's hint that the poor woman had been up against everything from sexism to anti-Semitism.

I decided to leave that section be.

SUSIE ORBACH

Sex and the Therapist:
Views of the Couch

MArilyn K. Yee/*The New York Times*

Susie Orbach, the therapist who treated Diana, Princess of Wales, for her eating disorders; the founder of the Women's Therapy Center of London; a former columnist for *The Guardian;* a visiting professor at the London School of Economics; and the author of 1978 bestseller *Fat Is a Feminist Issue*—is, aside from Sigmund Freud, probably the most famous psychotherapist to have ever set up couch in Britain.

On a recent visit to New York, Dr. Orbach, 53, talked about her eighth book, *The Impossibility of Sex: Stories of the Intimate Relationship Between Therapist and Patient* (Scribner).

___ *In* The Impossibility of Sex, *you own up to a great emotional involvement with your patients. Is this usual within your profession?*

Entirely usual. But analysts have a lot of different methods for dealing with it. Depending on their school, they'll say, "this is about my personal engagement."

Or, "this is what is done to me by the patient and it has nothing to do with me." Or, "this patient is disturbing me, I'll have to send him to another therapist." At the end of the day, though, we know there are no human relationships where we do not disturb one another. And that's true of the analytic relationship, too.

___ *In the book, you admit to feeling sexually stimulated by a patient named Adam, who kept trying to seduce you. Are you telling secrets out of school?*

I don't think there's any shame about that. The point is, How does one respond to the various emotional invitations of one's patients? Every person has an idiom. Some patients invite me in with their intellect. This particular man was a compulsive fornicator, and he had no other way to maintain a relationship unless he was involved in seduction. To get into his heart and head, a piece of me had to surrender to that seduction to understand what that was about for him and to understand how incredibly painful, barren, and problematic it was.

— *One gets the feeling you see psychoanalyst and patient as a kind of emotional couple.*

I see patient and therapist working together, creating a relationship in which you are lifting the impediments to intimacy. And you're often doing that within the relationship you have together. So in that sense you are a couple. But it's a very specialized type of couple—just like therapy is a very specialized type of relationship.

— *The Princess of Wales went to you for several years, did she not?*

I really don't know what I can say. . . .

— *When Diana gave her famous 1995 televised interview where she confessed to infidelity, eating disorders, and the fact that there were "three of us in this marriage," she was accused by several writers of talking "psychobabble." How did you feel about that sort of critique?*

You know, at the time, I was accused of being the scriptwriter, the person who put these words in her mouth. I found the accusation extraordinary. It's such a complete misunderstanding of the therapeutic process, as though therapists were Svengalis.

The truth is, we don't have an easy language for emotional life. That's why we have writers. They are able to find a way to say things with subtlety. So when a phrase rings true to somebody and then gets repeated, it becomes, all of a sudden, psychobabble. I think the problem with describing emotional life is that we have to stretch ourselves to find the words to say something that is refreshing, accurate, authentic.

— *After Diana's death, many commentators began looking at her life as a feminist parable. Did you?*

Do I think the transformation of a socially privileged but emotionally neglected child to a fairy tale princess, to a woman with a problematic marriage who turns around a sense of victimization to become a real activist in the world and fighter for people's rights as a feminist parable? Sure!

But is that the story of Diana? That's one story of Diana, isn't it?

— *Are you implying more to Diana's story than has been told?*

I think of hers as the story of someone who was vulnerable. And if that can be elevated into part of the feminist parable—that one has both strengths and vulnerabilities—I think that would be a welcome development to our thinking about women.

— *Your highly influential 1978 book,* Fat Is a Feminist Issue, *posited the idea that some women had eating disorders because they had been under-nurtured by their mothers. Do you still believe that?*

If I were writing that book today, I wouldn't write it in the same way. But I think that there still is a way in which certain aspects of female development are seriously unaddressed.

I think today women are told that they can conquer the world, but I don't think they are given the emotional equipment to feel safe inside themselves. And food is the first relationship. If we look at mothers' feeding of children, well, the situation is profoundly worse than it was in the 1970s. We now have two generations of women who have been under tremendous assault about: "You can go out there and be in the world, but don't forget to look gorgeous and slim. And don't forget, you should eat this and not that, and you should go to the gym." As people felt less effective collectively, politically, we've been offered our bodies as the site of change.

— *So the prevailing idea, in your view, now is, You can't change the world, but you can sculpt your body?*

I don't think that's explicit in people's minds. But I do think that people, in the face of feeling unhappy about their capacity to be effective, have found an area where they can feel they are. I think you can see progress in Western countries in terms of, certainly, certain classes of women's aspirations having been struggled with. It's sort of like there's been a very interesting price, which is an intensification in the breeding of the bodily insecurity of women.

— *Back to Diana. How did you personally deal with her death?*

I'm not really sure I can answer that. I mean, it's shocking when somebody you know and who is in the public eye has died. Mainly, I spent the day warding off journalists.

— *Did the harassment from the press directed at you help you identify with Diana?*

I have never admitted that Diana was my patient, even though it's in the public record. It is not something a therapist can say.

Of course, when your house is surrounded, and you have a clinical practice to deal with, the other people you are seeing have a lot to deal with they didn't bargain for. They are being invaded themselves. I mean, finding pictures of us in the newspapers on a daily basis. It was not nice.

— *In closing, your new book consists of eight composite case histories of psychoanalytic patients. Many therapists, including Sigmund Freud, wrote in the form of case histories. Why should readers believe veracity of a form that hovers somewhere between fiction and nonfiction?*

My book is about emotional truth. Why should you believe me? You don't have to. The question is, Is there anything there of use to you? Any time a therapist tells a story, they are being so selective. They are privileging one bit of information over another. In the end, it isn't really the patient's story anyway. It's the story the therapist makes of the patient's.

With the book, I was trying to say, "This is the therapist's experience." If you look from the point of view of the therapist—the scholarship, the learning, the knowledge, this heart—what I've written has a certain truth. *June 6, 2000*

Postscript

Susie Orbach is still doing what's she always done: seeing private patients and campaigning for public policy efforts aimed at discouraging the cult of thinness in women.

When I last heard from her in February of 2001, Dr. Orbach wrote from London that: "Recently, I have been a consultant and a keynote speaker to both the United Kingdom and the Viennese governments in their initiatives on eating and body image problems. They have (finally) realized that this is a major social and health problem.

"I am co-founder of two organizations making waves in Britain: the Campaign for Emotional Literacy . . . and the Psychologists and Counselors for Social Responsibility."

RUDOLPH E. TANZI

A Gene Detective on
the Trail of Alzheimer's

Rick Friedman

Hull, Massachusetts

On a recent frozen afternoon, **Rudolph E. Tanzi,** a 42-year-old neurogeneticist, was lounging at his beachfront home here near Boston, fingering the new book *Decoding Darkness: The Search for the Genetic Causes of Alzheimer's Disease,* which he wrote with journalist Ann B. Parson.

"It's a real kick for a scientist to have a trade book in the stores," said Dr. Tanzi, who trained at Harvard and is now a professor of neurology at its medical school and the director of the Genetics and Aging Unit of Massachusetts General Hospital.

"Sometimes, I go down to Borders and it blows my mind to see my book sitting there next to those of well-known authors."

Dr. Tanzi is particularly well qualified to write this book, a history of Alzheimer's disease and of the scientific battle to locate that dementia's cause and cure. In 1986, he was among the first investigators to discover the APP gene, which would later be identified as the first known Alzheimer's gene. Seven years later, he was the lead investigator on the identification of the gene that causes Wilson's disease, another neurological disorder. Then, in 1995, he was part of a team that found two more genes associated with Alzheimer's.

We talk over coffee.

—— *How close are we to an effective treatment for Alzheimer's disease?*

I wouldn't be surprised if five years from now we have a pretty effective drug that can slow the disease down enough so that it will be preventable in those at risk and significantly slow down the deterioration of people who already have it.

What we have learned is that all forms of Alzheimer's disease involve a common event: the excess of accumulation in the brain of that very nasty sticky substance, beta-amyloid. It gums up the space between the nerve cells, is toxic to nerve cells and interferes with the normal function of the brain.

Moreover, we now know that there are some defective genes that cause the disease very early in life and that increase the production of the amyloid. Other genes involve the more common late-onset Alzheimer's in patients 60 years and older, and affect the ability to clear the beta-amyloid from the brain.

— *How will the new medicines currently on the drawing boards attack Alzheimer's?*

The new medicines hit from both sides. One trial being conducted right now involves slowing down the production of the beta-amyloid. Two others are looking at how to promote the clearance of the amyloid already deposited in the brain. There is this idea around about a vaccine, where you actually inject the amyloid so that you have an immune response to it, and that promotes the clearance of the amyloid. This crazy idea seems to be working. And a company, Prana, which I have an interest in, has a clearance strategy based on a compound that sucks metals out of the amyloid deposits. If you suck copper and zinc out the amyloids, they fall apart.

— *Tell us, How did Alzheimer's become your life's work?*

I got into it because I wanted to be the first to build a complete map of a human chromosome. So I chose the smallest chromosome to study, Chromosome 21. My work on that brought me to studying Down's syndrome because those patients have an extra copy of that particular chromosome. Down's syndrome patients also get Alzheimer's pathology. So I had picked this little chromosome and without knowing it, I was studying Down's syndrome, and that led me to wonder, maybe there's an Alzheimer's gene on Chromosome 21.

Why else would all these Down's syndrome people who have this extra copy of this chromosome get Alzheimer's by middle age? And then, working backward from the composition of the beta-amyloid, I found the gene that makes it, and then I found where it was mapped in the genome. And guess what? It was on Chromosome 21!

— *While you were doing your research, your own grandmother developed Alzheimer's. What did that do to you?*

I was able to see the worst side of the disease. She became very anti-social, paranoid, aggressive. This is my grandmother! She used to make me homemade pasta and hold me on her lap, and now she's becoming this scared, paranoid, aggressive person. It was shocking.

— *Did her illness motivate your research?*

It motivated me to work harder. But I really think that what drives scientific discovery is the quest to solve puzzles. What drove my research is that same kind of feeling that doesn't permit you put *The New York Times* crossword puzzle down once you've started working on it. It's getting involved with a puzzle day and night wondering in the shower: What are the missing pieces and how do they fit together? It's just a pure obsessive, almost pathological curiosity that drives research.

— *There are two competing theories in the scientific community about the causes of Alzheimer's, are there not?*

There were two. One was: Is it the amyloid lesions? The other was: Is it the lesions that are seen in the nerve cells as they die—the tangles? These tangles are the twisted abnormal pretzel-like filaments that form inside of the nerve cells as they die. There are now overwhelming data showing that tangles occur in dying nerve cells following the accumulation of amyloids. So both theories are right. But if you choose to treat the tangles, which are really end-stage lesions, what you are doing there is letting everything go to hell and at the very last minute, coming in trying to save the system.

— *How intense is the rivalry between proponents of the theories?*

It was intense. But, as more research got done, we started to realize that the tangles are more an end-stage event and the accumulation of the amyloid fibrils happens at an earlier stage. That's important because if you want to nip things in the bud, you get them at the earliest stage.

Another thing, the tangles people eventually were satiated because they got a disease of their own to study. There's another form of dementia, frontal temporal dementia. The tangles researchers have found mutations in the gene called tau, which produces the material in the tangles. So there you don't have to have any amyloid but you have mutation in the tau gene, and you get the tangles and then you get a dementia, though not Alzheimer's.

— *You are divorced. Did your research kill your marriage?*

It surely ruined my marriage. My ex-wife and I are still friendly, but you know, I was working 12 to 15 hours a day. When I was discovering the amyloid gene back in 1986, I worked all night Christmas Eve, the entire day of Christmas, all of New Year's Eve and all of the New Year. It was obsession. Like I was so close.

"This could be it," I kept thinking. "What I'm working on could be the cause of all Alzheimer's!"

— *In the late 1980s, a lot of people went into studying AIDS because it was a terrific way to make a reputation and a career. Is that happening now with Alzheimer's?*

Well, I think now people are doing Alzheimer's because it's a good career move. When I started working in the area, there were a handful of people doing it. I remember when I took my neurobiology of disease course at Harvard Medical School, and each student got to choose a disease that they wanted to study for a term project.

Alzheimer's was the disease no one wanted and it was picked last. And that was 1984. Today, it would be the first one to go.

— *What is your take on the drug Aricept, which is widely used today to treat Alzheimer's patients?*

It's better than nothing. It helps an Alzheimer's patient to make the best use of the nerve cells that are still surviving in the brain. Unfortunately, after a certain amount of time, the effects can start to wear off. As useful as it is—Aricept is—it is still just a Band-Aid on a gushing wound. Right now, we just don't have anything else.

— *You are also a pianist. Last year, your recording, "Lost Sun," was one of the top selling CDs on MP3.com. Do you ever regret not having become a rock star?*

No, because science has filled that void for me. As a child I dreamed of traveling the world, playing my music. Instead, I travel the world presenting my findings. The musician's life is a rocky one. In science, if you do something good, you'll eventually get credit for it, but in music, you can be the best in the world, and it's still a crap shoot whether anyone will hear what you've done.

December 5, 2000

Postscript

Rudi Tanzi and I last spoke, appropriately enough, on Ronald Reagan's 90th birthday.

He was about to have a paper published in *Science,* in which his and two other investigative groups reported finding some evidence that a major Alzheimer's gene may exist on human Chromosome 10.

Dr. Tanzi was, as usual, breathless and excited about all that was happening in the Alzheimer's world. "Taking all three reports together, there is reason to believe," he said, "that there is an as-of-yet unidentified gene on Chromosome 10 with Alzheimer's. Even though we don't know what it is yet, the genetic data that led to its discovery has now triggered an intense race to isolate that gene. My guess is when all is said and done, there may be five major ones (Alzheimer's genes) found.

"For the genes that we have studied already, we (the scientific community) are making great progress in identifying new drugs based on what we are learning; specifically, we are branching out to looking into the roles of calcium and cholesterol in the Alzheimer's disease process. I think it's going to turn out that too much cholesterol doesn't alone predispose you to heart disease but to Alzheimer's too. We are also seeing that if calcium isn't regulated correctly, it seems to be correlated to Alzheimer's."

POLLY MATZINGER

Blazing an Unconventional Trail
to a New Theory of Immunity

Marty Katz

Polly Matzinger, 50, a tenured medical researcher at the National Institutes of Health, is turning the world of immunology upside down with her radical theory that the human immune system is activated not by alien substances in the body, but by dangerous ones, whether "self" or "alien"—the Danger Theory of immunology. She sat in her Bethesda, Maryland, garden recently, explaining the evolution of her ideas.

___ *As a child, what kind of work did you think you would be doing when you grew up?*

A jockey, a seeing-eye dog trainer, a composer. I grew up in a very creative, original family. My mother, a French ex-nun, is a potter. My father, a Dutch World War II resistance fighter who was imprisoned in Dachau, is a painter and carpenter. My sister is an artist, and my brother is a mechanic and rock musician. I sing, play piano and bass, and compose music. But I don't sing or play very well, and the music I write is dead-boring. The only thing I've found that I'm really creative in . . . is science. It wasn't an art form I expected to find, but then, my family has always blazed their own trails.

___ *How did you find science?*

I was a cocktail waitress in Davis, California, in 1972. I'd been in and out of school and done all kinds of jobs—jazz musician, lab technician, problem-dog trainer, and Playboy bunny, which was, incidentally, a great job. But everything seemed to get boring after a while, so I decided to make a career of waitressing and save the days for reading, composing, and working with animals.

One day, two professors from the University of California at Davis who regularly came in for beer, started talking about animal mimicry, and I asked them, "Why has no animal ever mimicked a skunk?" Professor Robert Swampy Schwab, who was chair of Wildlife and Fisheries, was

191

floored. He decided that this question-asking waitress should become a scientist. For nine months, he came to the bar and brought me all kinds of scientific articles. And he convinced me that science never gets boring. At his urging, I applied to graduate school. I started in 1974, got a belated BA in '76 and a Ph.D. in '79. I owe that man my life.

___ *When did you begin questioning the ideas that are the bedrock of immunological theory?*

In graduate school, we learned that the immune system fights anything that isn't part of our bodies. But that didn't make sense to me. I wondered why mothers didn't reject their fetuses, why we didn't reject the food we eat, or the stuff in the air we breathe. But my professors all said, "don't worry about it."

So I stopped worrying. This is a cowardly habit that we scientists can fall into. If we really can't answer a question, we sometimes stop asking it. Ten years later, I came to the National Institutes of Health and met a brilliant young oncologist, Ephraim Fuchs. He wondered about these questions, too. And we began thrashing them out.

___ *How does your Danger Model differ from the standard Self/Nonself Model of the immune system?*

It isn't really insurrectionary—it's just a different way of looking at things. Let me use an analogy to explain it. Imagine a community in which the police accept anyone they met during elementary school and kill any new migrant. That's the Self/Nonself Model.

In the Danger Model, tourists and immigrants are accepted, until they start breaking windows. Only then do the police move to eliminate them. In fact, it doesn't matter if the window breaker is a foreigner or a member of the community. That kind of behavior is considered unacceptable, and the destructive individual is removed.

The community police are the white blood cells of the immune system. The Self/Nonself Model says that they kill anything that enters the body after an early training period in which "self" is learned.

In the Danger Model, the police wander around, waiting for an alarm signaling that something is doing damage. If an immigrant enters without doing damage, the white cells simply continue to wander, and after a while, the harmless immigrant becomes part of the community.

How did you arrive at your alternate model for the immune system?

It didn't happen in a day. There were two difficult parts. First Ephraim and I took a year to realize that a truly useful immune system would fight dangerous things and ignore harmless ones. But we couldn't figure out how to tell the difference. One day, while I was in the bath, the answer just popped out! It was, "things that are dangerous do damage. No damage, no danger." It seems really simple, doesn't it? But it took us a long time.

Next we had to figure out how the white blood cells learn about the damage. That thought also arrived oddly. Annie, my border collie, and I were watching over some sheep grazing in a field. Suddenly, something moved in the woods, startling the sheep, and Annie jumped up to protect them. It hit me that something was amiss. Well, almost every organ of the body has a few sentinels—called dendritic cells—that are like sleeping sheep dogs. Alarm signals from injured cells could wake them up and alert the immune system.

What is the value of your theory of the immune system?

Besides the beauty of it? Well, first, it explains many things the old model couldn't, like why mothers don't reject their fetuses. They don't do damage. Second, there are some important practical implications. For example, the old model assumed that very young babies would be really hard to vaccinate because their immune systems were busy learning who the "self" was. Then we published a paper showing how to vaccinate one-day-old baby mice. Now there's a lot more hope about early vaccination.

Another hopeful area is cancer treatment. I really believe we can use vaccination to cure perhaps 80 percent of all cancers. Though there's a lot of research in this area, it isn't working very well. The Danger Model predicts that some simple changes could make anti-cancer vaccinations very effective.

But there is a lot of resistance to making those simple changes, based on your theory, isn't there?

The resistance isn't really to the changes. It's to the Danger Model itself. People grew up with a particular view of how immunity works. When I say, "That's not how it works, please accept that for a moment and you'll see we could do something effective against tumors," they are unable to try. It's partly fair. The Danger Model is new, and to experiment on people, you need to have a sound basis for the treatment. Still, I don't understand why they won't try it on cancer in animals.

— *What's the next big question in immunity?*

Once we know what activates it, we next have to learn how to make the right kind of response. Viruses aren't cleared the same way as worms and right now nobody has a clue how the immune system knows the difference. There's not even a theory. My lab is just starting to move in that direction, and I'm thinking hard about it.

— *I'm told you dislike the way science and technology have become hyphenated terms. Why?*

Because they are very different. Science is more like art, and true scientists are more like artists. Except that we don't have to starve in garrets because governments and foundations pay us. Technology is about vaccines and plastics and drugs and things that work in the world. Science is about describing nature, and so is art: We're painting nature.

— *Do you think the scientific world is too solemn?*

Oh, no. Not true science. It's art. Actually, it's a sandbox, and scientists get to play all of our lives.
June 16, 1998

Postscript

Polly Matzinger, head of the section on T-cell tolerance and memory at the National Institute of Allergy and Infectious Diseases' Laboratory of Cellular and Molecular Immunology, probably has the most interesting curriculum vitae in the entire world of American science.

It begins, intriguingly, with her activities from 1965 through 1974—jazz musician, dog trainer, carpenter, Playboy bunny, waitress, student (music and biology), Scientologist. The résumé ends with an equally fascinating list of recent professional papers she's coauthored with colleagues: "The maternal immune system's interaction with migrating fetal cells," and "On the lifespan of virgin T lymphocytes," and "Tolerance or immunity: Opposite outcomes of microchimerism from skin-grafts."

The last time I spoke with Polly Matzinger, which was April 2001, she was very much trying to reconcile a life of varied interests. The big change, for her, was the professional acceptance her theories were finally winning. "Things are really good right now. People are beginning to listen," she told me. "Data that twenty years ago people would have thrown out because they had no explanation for it, they find fits with my theory and so, now, they don't throw it out. They go and study it. It's really wonderful."

Matzinger's current project involves "working on ways to treat childhood leukemia immunologically."

The personal Polly has two new dogs in her life. That's in addition to the other two she keeps. "Lily is from the last litter of the world's herding champion," says Matzinger proudly. "It's clear she has an instinct about sheep that is remarkable for her age. Like her mother, she can push a ram, and she can also be very gentle with a lamb."

Less lamb-like is the other new dog, Charlie, the son of Roy, her wonderful canine patriarch. "He's a young teenaged male and rather teenaged male–like, not focused yet. He's a lover," declares Matzinger.

BENJAMIN S. CARSON

A Pioneer at a Frontier: The Brain of a Child

Marty Katz

Baltimore

It was 7:45 on a recent morning, and **Benjamin S. Carson,** professor of neurosurgery and director of pediatric neurosurgery at the Johns Hopkins Hospital, was sitting in his office here planning his day.

Dr. Carson, who performs three to five operations daily, has an international reputation as the man to call for tough pediatric neurosurgical problems.

Among his specialties, Dr. Carson separates conjoined (or Siamese) twins, performs hemispherectomies in which half the brain is removed from children who have extreme seizures, and performs craniofacial work on youngsters with congenital deformities.

If the medicine that Dr. Carson practices is heroic, so is his personal story. A 49-year-old physician, he grew up fatherless in Detroit. For most of elementary school, he was a problem child with miserable grades. But his mother, Sonja, a domestic worker, pushed young Benjamin to read books and eventually turn his grades around.

By the time he was ready for college, Yale was offering him scholarships. Medical school at the University of Michigan followed.

"There is a lot of personal satisfaction," Dr. Carson admits, "in coming this far from an environment where no one thought you could achieve."

— *What was your first reaction in 1997 when you were first asked to separate Luka and Joseph Banda, 11-month-old Siamese twins, who were joined at the head?*

I had mixed feelings about doing the operation.

I was in South Africa to receive an honorary degree at the Medical University of South Africa, and I was asked there if I might examine the twins.

Now, my previous experience separating conjoined twins in South Africa had not been pleasant. The Makwaeba sisters, whom we had separated some years earlier, turned out to be completely symbiotic.

We spent 20 hours separating them. But in the end, one of them had all the cardiac function and the other, all the renal. So the operation was technically successful, but the twins died. That really took a toll on me. I thought, "Why do this again, here?"

But when I saw the Bandas, lying there with their mother, playing, so cute, I became committed to them. I thought, "How can you not do everything to give them a normal life?"

In the end, the Bandas became the first Type 2 craniopagus twins (joined at the head and facing in opposite directions) ever separated with both surviving and both being neurologically normal. It was a triumph. It was tremendous.

— *Did the operation to separate the Banda twins prove to be difficult?*

Incredibly difficult. It was 28 hours of surgery. What helped a lot was that months before the actual operation, the experimental radiology division here at Hopkins asked me to look at a new computer they had that does 3-D virtual reality images of anything you program into it. I wasn't sure, at first, how such a computer might be useful to me. But when I returned from Africa, after my initial examination of the Bandas, it hit me: this would be great for separating Siamese twins!

One of the reasons surgeons have so much trouble separating Siamese twins is that nobody gets to do many of them. On the table, the anatomy is so different from normal that you're constantly trying to figure out, "Can I cut this? Does this wire lead to what?" It's like trying to defuse a bomb.

So when I got back from seeing the Bandas, I had their African doctors program all their information—CT scans, angiograms, MRIs—onto disks, and we loaded all of that onto the computer.

Here in Baltimore, I was able to have images of the twins' heads right there in front of me. It was beautiful because I was actually able to practice the operation. I could go from one side to another. I could actually go inside the blood vessel and see where I was going.

One of the exciting new things in neurosurgery is computer technology. It takes a lot of the guessing out of what you're doing. When you can practice going precisely to a place without poking here and there, the exposure of the patient decreases significantly, the morbidity decreases. Ultimately, the cost decreases because complications go down.

— *What do you feel when you're digging around in someone's brain?*

I'm constantly awed by it. Every time I am looking into the depths of somebody's brain, I'm thinking, "This is what makes a person who they are. That structure contains memories. Everything that they've ever experienced is right in there."

— *What happens to time when you're operating?*

It disappears. A six-hour operation can go by in what seems like a half hour. You're concentrating at such a high level.

— *With your heavy patient load, do you get to know any of them personally?*

Oh, absolutely. That's important. Getting emotionally involved, I think, is a good thing. Of course, when there's a bad outcome, it's difficult.

Just a few weeks ago, I cried right in the operating room because the patient was brain dead. I had grown so close to that family that I had a picture of the little boy in my pocket. And yet, we had done everything we could possibly do. The family knew that. The parents were in no way bitter. There were hugs all around. It really is difficult.

By the same token, I've never spent a day in court.

— *You were an underachiever as a child?*

Yes. I had a low opinion of myself like so many young people today.

When I thought I was stupid, I acted like a stupid person. And when I thought I was smart, I acted like a smart person and achieved like a smart person.

I was fortunate, in that I had a mother who believed in me and kept telling me I was smart.

You know, even as late as my first year in medical school, my faculty adviser advised me to drop out. He said I wasn't medical school material.

Interestingly, I told that story last June when I was the commencement speaker at my medical school.

— *When did you realize that neurosurgery was your calling?*

I first wanted to be a psychiatrist. I decided against that in medical school when I discovered that psychiatrists didn't, in reality, do what they did on TV. So I had good eye-hand coordination, I was a very, very careful person, I loved to dissect things, I could think in three dimensions. Putting all that together, I decided on surgery.

— *You are probably one of the few African-American neurosurgeons in the country. It must be a little annoying to always be pointed out as the exceptional guy who made it big.*

You're right. And I'm not the exception. My brother has done very well also. He is the manager of the aircraft landing division of Allied Signal, which recently merged with Honeywell. I know tons and tons of people who grew up in environments like mine who are doing fantastic things intellectually and nobody knows about them.

— *Tell us about the Carson Scholars.*

When I'd go to schools to speak, I saw all these trophies for sports and that the kids who were in the National Honor Society just got a kind of pat on the head. They were derided as nerds and Poindexters. I thought, "This is crazy. These are guys who will keep us number 1, not the guy who can do the 25-foot jump shot."

So my wife, Candy, and I decided we'd take $500,000 of our own money and start giving scholarships to students to achieve at the highest academic levels and also who demonstrated humanistic qualities. The money is invested. When they go off to a college, the money gets transferred there. The school also gets a big trophy, a lot of local publicity; they get a medal to wear, and they go to a big banquet.

— *The writer John Horgan, in his book* The Undiscovered Mind, *claims we know next to nothing about the way the brain really works. Do you agree?*

I agree with him 100 percent. It's one of the things that attracted to me to neuroscience. I thought, "Here's a field where you can become a real expert pretty quickly because nobody knows anything."
January 4, 2000

Postscript

Dr. Carson continues in his scientifically innovative and socially concerned style of medicine. In addition to his ongoing work with children, he has, according to Wendell Mullins, a public relations officer at Johns Hopkins, begun "treating an increasing number of adult patients with trigeminal neuralgia, a nerve disorder that causes extreme facial pain and is an underdiagnosed and difficult disorder to treat. He has achieved an 80 to 90 percent rate of success for more than 100 adults he sees annually."

POLICY MAKERS/
POLICY CHANGERS

RITA COLWELL

"Always, Always, Going Against the Norm"

Marty Katz

A year ago, when **Rita Colwell** was sworn in as director of the National Science Foundation, new ground in science and politics was broken. Dr. Colwell, a microbiologist now 64 years old, became the first woman and the first biological scientist to head the federal agency, whose $3.6 billion budget provides financing for basic nonmedical research in science and engineering.

Before accepting the foundation post, Dr. Colwell was president of the University of Maryland's Biotechnology Institute and one of the world's major experts on cholera.

She spoke during a break in a recent meeting of the American Association for the Advancement of Science in Anaheim, California.

___ *You went to Purdue University during the early 1950s, the time when the ideology that held that women were best off as housewives and mothers was burning strongest. What was it like to be a woman with scientific ambitions during that era?*

It felt peculiar wanting to be a scientist when everyone around me majored in home economics. On the other hand, I had encouraging teachers. For instance, at Purdue, I took a course with a bacteriologist named Dorothy Powelson. A woman bacteriologist in those days was unbelievably rare. Her approach was, "Look under the microscope and what do you see?" She got me hooked.

My goal, originally, was to go to medical school. But in the last semester of my senior year, I met this handsome 6-foot-2 physical chemistry graduate student, Jack Colwell, and after one date, we decided to get married. The marriage, incidentally, has lasted for 40-plus years. So I think it was a good decision. But once I made it, medical school was out.

So I began looking for a bacteriology fellowship. I went to the chairman of the department of bacteriology and told him of my new plans. "We don't waste fellowships on women," he declared.

The day was saved by my adviser, Dr. Alan Burdick, a geneticist, and he said: "O.K., their loss is our gain. Would you like a fellowship to study genetics?" This was the foundation of my molecular biology career.

— *Did you encounter much prejudice during those years?*

Oh, yes. In 1961, my husband was awarded a fellowship in Canada. When I applied for one, too, the Canadian National Research Council rejected me because of their "anti-nepotism" rule, which precluded offering fellowships to husbands and wives. Eventually, I obtained some funds from the National Science Foundation and was offered laboratory space in Ottawa. The University of Washington, where I'd received my doctorate, then appointed me a research assistant professor and granted me a leave of absence. So in this complicated way, I was able to be with my husband and still do my research.

You were always, always, going against the norm, the dogma. And that does wear on a person after a while.

— *Is the path still full of obstacles?*

They are there, sometimes. Especially when you're heading organizations that used to be run by men. The academic world is still a closed circle. I try to work around that.

For instance, sometimes at meetings, a woman will suggest something and there's no response afterward. Then, a male colleague says exactly what you said and they say, "By God, that's a good thing to do!" I've learned to seed the conversation quietly with my ideas, and then they get adopted. You don't get the credit, but if you're concerned about credit, nothing gets done.

— *About your area of research, cholera. Why was your discovery—that the cholera bacterium exists in a dormant stage in most of the world's water—a benchmark in the control of the disease?*

Because before our findings were published, the medical community always believed cholera was transmitted person to person, that it had a human host or reservoir. We showed it exists in the environment, on plankton, and in a dormant stage between epidemics. The implication of that is that when the plankton populations bloom in the spring and again in the fall in countries like Bangladesh, the bacterial numbers go way up along with the increase in numbers in plankton. So if the water that is used directly for drinking has large numbers of cholera bacteria in it, there may then be enough bacteria present to cause the disease.

— *One idea you've come up with for cholera control in Bangladesh is getting people to filter their drinking water by pouring it through layers of sari cloth. How did such a simple notion come to you?*

I thought if one could remove the zooplankton—the copepod on the zoo plankton serve as host to the cholera bacteria in drinking water—we could go a long way toward curbing the disease.

The problem was that sophisticated water filters were too expensive for Bangladesh, one of the least wealthy countries on earth. So the thought was, "What could be used as a filter that exists in everyday life?"

The answer was cloth—sari cloth—which even the poorest of the poor have.

— *Has your method gained acceptance?*

Where people use it, yes. Recently, I was in Bangladesh with a crew from Maryland Public Television, and we went to a village where we asked a woman if she'd gather water for drinking.

She did that and she could see plankton and various larval stages swimming about the unfiltered water. "The cause of cholera is on the particles in the water," I explained. We then showed her how to use the sari cloth as a filter. The filtered water wasn't crystal clear, but it was clearer than the unfiltered water. There was nothing swimming in it.

After that, the cameraman wanted a shot of her going again to collect water and drinking it, and she said, no she wasn't going to ever drink unstrained water anymore.

That woman "got it," in a minute!

— *Let's talk about some of the areas of research the National Science Foundation supports. Do you think now that the cold war is over that the United States should be financing more research in Antarctica?*

Oh, absolutely. The things we can learn about atmospherics science, physics, astronomy are done extraordinarily well at the South Pole. It's the least contaminated area in the world.

__ *Do you have an opinion about human embryo research?*

I agree with Harold Varmus, the head of the National Institutes of Health, who said that human stem cell research is very important. It's an opportunity for us to learn how human cells differentiate and the potential is so great for developing a capacity for producing organs. I think the value to humanity, plus the wonderful information it can bring about life and development, is so very, very important. I do think it's very fundamental.

__ *Vice President Gore recently announced a $366 million a year information technology research initiative that the N.S.F. will be heavily involved in and that will finance new computer hardware and software research. Why can't the computer industry, which is doing quite well, finance this sort of thing itself?*

Because the basic software research requires this kind of investment. You need it in order to keep America at the cutting edge.

Let me just give you a vignette. I recently visited Seattle and the computer science department at the University of Washington, and I met incredibly bright young people who were soaring like eagles with their research. Afterward, I went to Microsoft and again met some incredibly bright people. But it was like watching eagles soaring within the aviary. They were netted in because they had a product-driven challenge.

The United States, as a nation, has got to be at the forefront of this technology. The competition is keen from other countries. When they put up a new university in China, the first structure they build is the computer sciences building. Our military, our medical institutions, need to go that next step for those new kinds of languages that will link up a thousand processors for the kinds of research we need to do.

__ *On a different subject, do you think physicists make great husbands?*

(Laughs) I think, yes. I know that anecdotes don't make statistics, but I was once on a panel for the National Institutes of Health and there were four prominent women there speaking. Afterward, we had lunch together, and it turned out that each one of us was married to a physicist.

We concluded that physicists have interesting work, a wide range of interests and are not overly concerned about their masculinity. Now obviously, you will find physicists who have none of these characteristics. But in general, we all felt this was so.

___ *So your advice to a bright woman looking for an appropriate man is: Go to a physics convention?*

I'm not quite sure it's that simple. What I think you have to do is find someone who is interested in you as a partner, not as a domestic. My husband is the most gentle and supportive soul I've met in this life. We've been married a long time and it's been a powerful partnership. He's been a superb father and much involved with our two daughters. Both of them are scientists, by the way. Alison, 36, is at Harvard.
February 16, 1999

Postscript

Interviewing the handful of women who started their careers in science during the 1950s is always fascinating. These are women who became scientists at a time when everything in society was working to channel their creativity toward the home and when the only kind of chemistry they might have experimented with, was in the kitchen.

The 1950s women who made their careers against incredible odds almost always have big stories to tell. Moreover, one rarely hears from them the cliché that women my own age and younger often recite: "If I made it, any woman can. There's no such thing as discrimination."

Dr. Colwell, marvelously unaffected and open, was a perfect example of 1950s pioneerism. I wished there had been room to print all of her interview which was full of examples of the kind of subtle and unsubtle discriminations she's battled all her life.

Surprisingly, in the interview with Dr. Colwell, I was interested (as is she) in more than her professional accomplishments, but also in her gorgeously happy personal life. And while it is true that I might not ask a male scientist about his marriage, I had met Jack Colwell and was impressed by his supportiveness. I thought their marriage was worth a question or two.

Well, the armies of the Politically Correct really came down on me for that one. My editors received a sheaf of stormy letters from academic women claiming that I had "trivialized" Dr. Colwell. I was a sexist. I was dumb.

Thus I was amused some weeks later to interview another pioneering woman, Eleanor Baum, and to discover that she too was married to a physicist! I know: Anecdotes make for bad science, but it seemed that the Colwell Thesis on Potential Husband Material might well be worth some enterprising sociologist's time and effort.

JEFF GETTY

Still Living With AIDS, and Endless Jokes About Bananas

Peter DaSilva

When one first meets **Jeff Getty,** 41, at his Japanese-style loft in Oakland, California, it is hard to think of him as a man who has lived with HIV and AIDS for 18 years. Mr. Getty gives off an air of vitality. His eyes gleam; he is muscular, articulate, and energetic. The only sign of AIDS is an infusion tube filled with antibiotics sticking out of his right arm, an emergency measure he needs to fend off a nasty lung infection. But Mr. Getty has learned to live with AIDS, even defeating death.

Two years ago, Getty made international headlines when he persuaded the Food and Drug Administration to permit him to have an experimental bone marrow transplant from a baboon. The graft did not take, but Mr. Getty believes that the radiation treatment he received as part of the transplant procedure may have extended his life, along with a succession of experimental therapies and his "warrior" attitude about the disease. At any rate, his transplant was controversial. And the aftermath, surprising.

___ *What happens to you, Jeff Getty, when you eat a banana in public?*

I can't do it. People go "urk, urk." They start laughing uncontrollably, scratching their underarms, and making embarrassing jokes like, "How are your friends at the zoo?"

After I participated in the experiment, the jokes really came hard. It began with gifts of stuffed monkey toys. Next, somebody made a tape of songs about monkeys. I got a banana holder as a gift. Then I went on the Tom Snyder show and he tried to hand me a banana. There I was, going, "Tom, don't try to hand that to me on camera." I thought all the ribbing would eventually stop—especially since the baboon bone marrow graft didn't take. But this went on for years. What I suspect it's all about is people alleviating their own discomfort with what I've done.

___ *And what exactly did you do that makes people so uncomfortable that they make banana jokes?*

Cross the border of human-centrism. I've come to think that when we do cross-species research, the people who participate are thought of as not necessarily 100 percent human afterward. There's this unconscious sense that the procedure has created something adulterated, a human-animal.

— *Are you a human-animal?*

Well, when you take baboon bone marrow and put it into somebody, that's a pretty strong statement. Bone marrow is the progenitor of all the cells of the body. It would follow that many of the cells of the person would become baboon cells—not just the immune cells, but also the red blood cells. After the procedure, the question is, "Is the person 100 percent human?" With something as fundamental as a bone marrow transplant, it's questionable.

— *Do you feel you are not 100 percent human?*

It's not an issue, I feel. My body rejected the bone marrow rather quickly. But most people don't know that. To this day, most people think I have baboon life within me.

You know, when you look at horror movies, you have to wonder: Why are they so fascinating, these films about werewolves and things that crawled out of the ocean part fish, part human? Human culture is full of fascinating mythology about centaurs and werewolves. And I have the feeling that what I've done is touch many people's, including many scientists', deep unconscious fear that xenografting means creating werewolves and monsters.

The thing that's so devastating because there are so many areas where xenografting is logical and necessary, is that this is a fear that many scientists won't admit to. You know, it's not "scientific" to be afraid you're making a monster.

— *How did you get to the idea of doing a baboon marrow transplant?*

I'm a member of "The Dobson Project," a west coast think tank on immunology that was founded by the activist community here. Once a year, activists and top immunologists meet for three days and look at ways to restore human immune systems. We design new protocols for this.

The baboon marrow transplant was one. Baboons have a natural resistance to HIV. So if one could get a baboon immune system to engraft temporarily into a human, the question is: Might it help that human being get rid of the virus? Could the baboon cells teach the human cells how to fight the virus?

We were hoping to create two immune systems functioning side by side, the human and the baboon—what is called "a chimera immune system." This has been done in animals quite a bit in research. You can't get one immune system to completely agree that the other is self and, thus, will not be rejected. You have to trick them into coexisting, side by side, instead of fighting each other.

— *How was your health at the time you wanted the baboon marrow transplant?*

Very, very bad. It was December of 1995, my worst year. I was failing all the new drugs and was close to death. So I thought, "What the hell, why not go into an experiment that had a long-shot chance of helping?" Honestly, I didn't expect to survive.

— *Weren't officials of the FDA extremely negative about the procedure?*

Negative? They put endless obstacles in our way. In the end, we were able to mobilize a good part of the activist community and the experiment went forward.

— *But scientists from the FDA and the Centers for Disease Control and Prevention were afraid you might be risking the transmission of new viruses from baboons to humans. Isn't this a reasonable concern?*

There's validity to being concerned about cross-species diseases, but it shouldn't stop the research.

The answer is: Control the experiments and monitor the people who are experimented upon. Any person who has had this kind of xenografting should assume that they might have some unknown viruses. They should be taught about safe sex and how to avoid transmitting the disease.

But interestingly, a lot of the people who work in animal husbandry and the veterinary sciences are exposed to all kinds of animal diseases and they are not being monitored.

The C.D.C. issued a report on 15 lab workers who they found monkey viruses in. Some of them had carried the viruses for 20 years. Now, those are people who should be monitored and taught about safe sex and condoms, but who's even suggesting it? My sense is that if this fear of animal diseases were a really valid thing, based on sound science, then people in the veterinary sciences should be working in spacesuits.

— *After the operation, you had lots of communications from the folks at the C.D.C.*

I'll say. Afterward, it seemed to me that the C.D.C. was hoping to find a baboon virus in me, that they were even disappointed when they didn't. Every time they'd discover a new baboon virus, they'd be on the phone saying, "We want more blood to see if you've got this one."

It was pretty gross sometimes. I felt I was carrying the future of cross-species research on my shoulders. As long as they didn't find any disease, the research could go forward. But if they found something, this very valuable possibility for humans might end.

— *There are people say you're being humano-centric by encouraging the use of animals as spare parts for humans. How do you answer them?*

Oh, you mean the animal rights people? Well, you know, People for the Ethical Treatment of Animals has called for all AIDS research to stop using animals, so it's hard to take them seriously. ACT UP, in fact, has demonstrated against them. Frankly, if my life is in danger or my child's, and a baboon heart would help, I would say, "Yes, I want to go forward with that."

— *At present, humans receive heart valves from pigs. Why isn't there the same kind of opposition to that?*

Well, pig valves are another thing, entirely. They don't carry viruses. And even then, I bet that people with pig valves get lots of pig jokes. I'm telling you right now: People who have pig valves on their hearts don't go around advertising it.

And people who get pig brain isolates, where they place cells for people with epilepsy and Parkinson's into the brain—you can just imagine what will happen there. They are going to be called "pig-headed."

And people are going to wonder if they are thinking right or thinking like a pig.

Listen, if I can't eat a banana in public, what's going to happen to people with pig implants in their brains?

As xenografting continues, we're going to see, I think, society having a hard time with it. We're going to see irrational fear of animal diseases and the shunning of people who've had animal transplants. It's going to be a long time till this procedure is as accepted as, let's say, human to human heart transplants have become. And think about that, 20 years, people were concerned about cross-racial organ transplants, and this is a much bigger border. And yet all science tells us, we're heading right for it because of the shrinking pool of donor organs available for humans and the growing technology that makes transplants possible.

___ *Do you have any regrets about doing the experiment?*

None whatsoever. I want to stand up for this.
October13, 1998

Postscript

The e-mail I received from Jeff Getty in February of 2001, three years after our interview, was disturbing. In addition to lots of news of his activist projects, this very long-term AIDS survivor wrote, "Yes, I'm still alive and fighting. . . . My health is poor these days and I have slowed down."

Pangs of guilt descended. . . . Like many others, I had come to see the wonders of medicine as saving so many with AIDS that I had stopped thinking of it as a lethal disease. In my own circle, I had friends who had been brought back from the very brink of death by the fantastic new anti-AIDS pharmaceuticals.

I'd been to Oakland twice since the interview and might well have stopped in and looked in on Mr. Getty. On both trips, time and work obligations had overwhelmed. I didn't do it. The guilt gnawed. Most journalists keep a strong professional distance from their subjects—and the reasons for that are valid—but if someone has touched me, I do try to stay in touch. Even though I had my questions about some cross-species transplantation, I admired Mr. Getty's bravery. His life force seemed enviable.

And so I phoned him—it was as much a personal as a professional call. Getty was wintering in Palm Springs, saving his frail lungs from the damp Bay Area climate. "I've been having a lot of opportunistic infections," he explained. "I'm talking all of the very latest experimental HIV drugs, and I'm only having limited luck. It is keeping me at a state of malaise. My life doesn't seem to be in danger, but I'm certainly not healthy either. It's the same old things one hears from AIDS people: 'I have my good days, I have my bad days.' Today is a **good** day."

On the policy side of things, Getty and his organization—which had changed its name from ACT UP to SURVIVING AIDS—had won some remarkable victories. "We've opened the door for people with HIV to get liver and kidney transplants," Getty reported happily. "It used to be near impossible for people with HIV to get on the transplant lists. We negotiated, put tremendous pressure on institutions that do this, and we also worked on the inside helping put together research teams and helped design protocols. It's now become an NIH national trial. Private insurance is beginning to pay. Within a year and a half, people with HIV won't be excluded anymore."

Mr. Getty paused and then said, "You know, Claudia, there's a guy down the block here in Palm Springs who I saw this morning, working in his garden. He had come to me and asked for help. He had HIV and needed a liver transplant. We helped him. Without our help, I doubt he would have gotten one . . . he's doing really well. He has a little boy. The reason I tell you is that it illustrates that there is somebody alive there—digging holes and planting plants and doing things that alive people do. And he wouldn't otherwise be there, if not for the work. And that's a really rewarding feeling."

STEPHEN E. STRAUS

Marty Katz

The open-faced man sitting behind an wooden table in his office at the government's National Insititutes of Health (NIH) is **Stephen E. Straus, 54**, a virologist and the holder of what may well be the most extraordinary job on the NIH's sprawling Bethesda, Maryland campus. For the past eighteen months, Dr. Straus has served as Director of the National Center for Complementary and Alternative Medicine (NCCAM), which means that Congress has voted him almost $90 million to study the usefulness of such popular nontraditional remedies as acupuncture, food supplements, homeopathy, and body manipulation. With 42 percent of all Americans—according to a study by Dr. David Eisenberg of Harvard—using various forms of alternative medicine, Straus's task is to find out what works and what doesn't.

Dr. Straus's own training is firmly rooted in the world of traditional science. He has degrees from M.I.T. and Columbia University College of Physicians and Surgeons. As a researcher, he's studied Lyme disease, AIDS/HIV, chronic fatigue syndrome, and the various forms of herpes infections. He's also served as chief of the Laboratory of Clinical Investigation at NIH's National Institute of Allergy and Infectious Diseases.

"When I was studying herpes," he smiles, "people would take me aside at parties and tell me their herpes stories. Now that I'm at NCCAM, they are plying me with stories about their experiences with St. John's wort and shark cartilage."

—— *After you took your new job here, a reporter for* Science *magazine wondered, "Why in the world would a respected researcher like Stephen Straus leave a topflight lab at the National Institutes of Health to run NIH's new National Center for Complementary and Alternative Medicine?" Okay, give us an answer to that question?*

I took it because I think the only way to change the dialogue on complementary and alternative medicine—what we call CAM—is to have a serious person here at the NIH, doing serious work, with serious funding. The fact is that Americans are using complementary and alternative therapies. If the public is spending billions of dollars on these things, they are either deluded *en masse*, or there is some communal wisdom they are expressing. I believe that the tools of science can provide very powerful answers on what they are doing.

___ *There are people on this research campus who are opposed to the very idea that something called the Center for Complementary and Alternative Medicine even exists. How do you handle them?*

There are probably fewer people who feel that way than you think. Frankly, there are people who are very vocal from both extremes. There are people from the alternative medical community who feel that the only reason we exist is to provide data for the government to get rid of them. On the other hand, there are people who believe that CAM therapies can't possibly work and even if did, they couldn't work any better than things delivered through conventional medicine.

I can assure you that before I took this job, I talked to the directors of lots of the institutes here to get a sense if they were as cynical about CAM as some people have suggested. They are not. They have not only been eager to collaborate with us, but for the first time to co-fund research projects.

___ *You have a significant budget: nearly $90 million per year. When you are deciding how to spend those dollars, what are your criteria?*

We have a complex process. My sense is that we must study first the things that are of greatest public health importance. If someone has a wonderful alternative treatment for hangnail, it's obviously not as important as an alternative to prevent Alzheimer's or to treat pancreatic cancer.

Our next criterion is, "Are these things of importance to the American people" as opposed to elsewhere. The third criterion has to do with where the opportunities are. There are thousands of different herbs and approaches from all over the world. We can't study them all. So we invest in studying things that are ready for a first serious look.

___ *One of your predecessors here, Dr. Joseph J. Jacobs, came under huge attack from traditionalists when he implied that medicine wasn't a com-*

pletely objective science in the same way that physics is. Why was this heresy? I always thought that medicine was a science and an art.

I've never met Joe Jacobs and I don't know that quote. I will say that medicine is both a science AND an art. And there's nothing wrong with saying we do not have received wisdom in all things. It depends on how you say them. To say that we do not have all the answers is not an excuse for (not) seeking them out. Dr. Jacobs was CAM's first director and the office had a completely different scope of authority and responsibility. Its budget was $2 million. We have forty-five times that. We have the authority of an independent NIH Center. Maybe I don't have as much need to be defensive and cautious because we have the tools to get what needs to be done DONE.

—— *Are you the medical equivalent of Nixon in China—a man with mainstream credentials who can do things that others cannot?*

Gee, that's an interesting analogy. We have to convince a wide spectrum of opinion that we are doing a good job. So far, my colleagues in science are glad that I'm doing this as opposed to someone less well credentialed. If the purpose here is to do this seriously, you need a serious scientist to do it. To the degree that I'm that person, it was a good decision. But we still have to prove ourselves. The people who are advocates of CAM, they've given enormous support to me—though not everyone.

—— *When you were a kid growing up in Brooklyn in the 1960s, did you dream of becoming a physician?*

No, I wanted to be an electrical engineer or a physicist. But when I got to my freshman year at M.I.T., I quickly discovered that the people who did physics were REALLY smart. Oh, I did fine in physics. But it was hard work for me. I realized it (physics training) was a formidable process and that people who excelled in it were wired a little differently from I.

Then, early in my sophomore year, I took introductory biology and I got the highest grade in the class. Hey, I had facility for this. I really loved it!!!

—— *You've had a career of being very much "where the action is" in biological research—studying, among other things, the herpes virus and chronic fatigue syndrome. On the latter, can you answer the million-dollar question: Does it exist?*

Of course, it does. But what is "it"? Is it a single disease? I don't think so. Is it caused by an infectious agent? Very unlikely. There isn't an infectious agent involved in sustaining the disease, as opposed to being a trigger for it. I think what we're calling "CFS" might be the common pathway of how our body expresses a series of assaults on it. Some might be physical stressors and some might be emotional.

What's important about CFS is: Many people get over it. Individuals who have it for many years lose hope. They then take on a series of maladaptive behaviors which sustain their illness because they become so focused and so phobic: They avoid exercise, disrupt their sleep patterns. It gets harder and harder for them to regain normalcy.

___ *Getting back to Complementary and Alternative Medicine, as one looks through a list of of studies this center has funded—St. John's wort, shark cartilage, ginkgo biloba—one senses that you are researching substances in current use and where there's been some anecdotal history of effectiveness. Am I right?*

Only in part. We invest in things where there's more than anecdotal proof. We looked into St. John's wort (an herb used for alleviating mild depression) because there were a couple of dozen trials of it, most of them small, and the analysis of all those studies has concluded that that is an area ripe for large clinical trials. The same is true for glucosamine chondroitin for acupuncture for certain kinds of pain disorders, for ginkgo biloba for preventing certain kinds of dementia.

Yes, we are very pragmatic. Some call that "picking the low hanging fruit." But there are people who have such a distrust of the research establishment that they think NIH is incapable of entertaining any serious research in this area. Thus, we need to show that we can answer some of these questions. Humans are already using these therapies and so we don't have to spend as much time examining what the molecules do in mice. We can go right to people in the research trials.

___ *You mentioned clinical trials on ginkgo biloba, which some people use as an Alzheimer's preventive. How are you studying its effectiveness?*

We have three thousand individuals taking at random either ginkgo or a placebo. We're doing it in partnership with the Mental Health, Neurology, Heart, and Aging Institutes. The prior published data is very encouraging.

But this is a flyer. If what we find is positive, we've got the first tool, however potent, to prevent the onset of dementia. If the results are negative, then this will have been the largest prospective study ever done on the first development of dementia in an otherwise healthy aging population. In either case, what we learn will be invaluable.

— *You've been heavily criticized for funding a study of the Nicolas Gonzalez protocol cancer therapy. He's into enemas, isn't he?*

Enemas, and a whole number of nutritional supplements. This is for an incurable cancer. The best chemotherapy for it affords an extra few months of life. The aggregate of anecdotes from Gonzalez—and that's all they are—is that in a handful of selected patients, the average survival is over eighteen months. I don't have any reason scientifically to feel that the Gonzalez regimen should work. On the other hand, here's something that addresses an incurable disease for which the best available therapy is poor. I think we should be willing to tolerate some discord and skepticism for the sake of getting a clear answer. If it doesn't work, we have a clear answer. If it does work, we will try to figure out why. Our study is funding Columbia University to run it, not Gonzalez. I have no problem justifying this one right now.

— *If you were seriously ill, would you do conventional or complementary medicine? Or perhaps both?*

You know the adage that a lawyer who takes care of himself has a fool for a client? Well, the same is true of a physician. I do think I am very positively inclined to what conventional medicine has achieved. I would do my homework because that's my nature. I doubt that I would reject out-of-hand any conventional therapy that holds promise. Would I gravitate toward other things as well? Probably. But I can't really put myself in that position.

— *Do you ever go for alternative therapies?*

No, I don't. I'm not an advocate for alternative medicine. I'm an advocate for science. I don't come in attached to some of these treatments. And this message has been discomforting to some who might have suspected that I'd be more accommodating to their beliefs. But I don't think I need that. It's not a part of my job.

April 3, 2001

Postscript

When I wrote Stephen Straus for an update on his activities since our interview, I felt a wee bit . . . embarrassed. It had only been a week since our interview had appeared in the *Times*. How much could have happened in his life since then? Had he perhaps been spotted by the press hanging out at a GNC store at a Maryland shopping mall? Had he taken up yoga?? Was he going for regular consultations with the Dalai Lama's personal physician??? All was possible. And one always hopes for good copy.

Here was his e-mailed response: "Well, since the article appeared only a week ago I have not yet been nominated for the Nobel Prize. . . . There is really very little to say, isn't there?

"What I can say is that I am gratified by the number of friends and colleagues and fellow medical students who read the article and contacted me about it. We all felt it captured well my enthusisam about meeting the challenges of exploring CAM approaches in a rigorous way."

RUSH HOLT

At Last, a Politician Who Knows Quantum Mechanics

Michael Geissinger

One of the upsets of the 2000 election was the defeat of the incumbent Republican, Michael Pappas, in New Jersey's 12th Congressional District by **Rush Holt,** who had never held elected office. Mr. Pappas was best known nationally as the Congressman who once sang, on the floor of the House of Representatives, a tribute he had composed to the special prosecutor: "Twinkle, twinkle, Kenneth Starr, now we know how brave you are."

Dr. Holt, 50, a physicist, was the assistant director of Princeton Plasma Physics Laboratory. He joined Representative Vernon J. Ehlers, Republican of Michigan, as one of two physicists serving in Congress.

"We're talking about starting a bipartisan physics caucus," Representative-elect Holt, a Democrat, joked in an interview at his mother's Washington apartment.

—— *In your campaign, you received contributions from 14 Nobel Prize winners. Did they support you because you're likely to put science issues onto the legislative agenda?*

I must tell you, we didn't try to recruit Nobel Prize winners to my campaign. But word went out in the community that I was running, and some scientists told others, and they gave. My campaign staff was very excited by these contributors, and they wrote them a letter and asked if it was O.K. to use their names. My staff picked up on that . . . just like they picked up on the fact that I'd been on *Jeopardy* six times.

—— *How much did you win?*

About $5,500 and a car. This was back in the 1970s when the categories were worth $10 and $20.

But back to my campaign: To be honest, science was not much of a factor in it. I would talk about what a scientist could bring to Congress, but issues of science did not seem to move most voters. Occasionally, I'd meet a voter who said he'd like to see someone in Congress who "really understood quantum mechanics."

But most people cared more about health care, Social Security, education—issues that they felt affected their daily lives, and I tried to address those.

— *Now that you've won, do you intend to be the science community's point man on the Hill?*

I can't help but be, yes.

— *Are you starting a trend, trading laboratory for legislature?*

With all due respect to lawyers, we need more diversity among legislators. I think more scientists should get involved in public policy. But many say: "Politics is dirty. There's too much compromising."

You know the first question everyone asks me is why would a scientist leave a perfectly good job at a major research lab to go into "that muck." But for me, politics wasn't such a big step. This is the first office that I've run for, but I worked on Capitol Hill in the early 1980s in the office of Congressman Bob Edgar as a science, defense, and education adviser. I worked at the State Department during the Bush Administration, doing arms control.

— *How do you think your father, the late Rush Holt Sr., a Senator from West Virginia, would have felt about your career change?*

I don't think he'd be particularly impressed or pleased with my election, although I showed some skill as a candidate. I think his satisfaction would come if he saw that I was really using the office to accomplish good for people.

My father passionately fought for good government. . . . Ever since my election, I'm getting letters from people in West Virginia who say, "Your father was a hero of mine." He's been dead 45 years, but he's still remembered because he took stands. In fact, he was defeated for re-election in 1940 because although elected as a New Dealer, he fought with Roosevelt on a number of issues.

Last week, during an orientation session the Democrats had for new members of Congress, one of the Congressional leaders said to us, the freshmen, "If you're not prepared to leave Congress over a principle, then you should leave now." I thought that was an interesting statement. So we'll see. . . .

___ *Why is it important that there be scientists in Congress?*

It's not that scientists have a corner on the truth, but a science background is important for understanding the limitations of some policies. Scientists are in a position to define what is possible. There are examples where policy makers promote programs that just are fallacious, that essentially are prohibited by the laws of science. Legislative calls for a space-based Star Wars system had some aspects of this.

___ *Speaker Newt Gingrich, who is retiring, is said to be a big science buff. Did he support science?*

One of the first things Gingrich did was to abolish the Office of Technology Assessment, an agency that advised Congress on science and technology, thus saying in effect, "We don't want scientists telling us what we don't want to hear." After that, I found it very hard to accept Gingrich's claim that he was the champion of science and technology.

___ *Would you like to see the office re-established?*

I don't know what the political possibilities on that are. But I do believe we have to increase the emphasis on science in the government. If there's a slogan on the wall, it should read, "It's productivity, stupid!"

The way our economy is able to provide for its people is through productivity and that comes from educational training and new ideas. The new ideas, primarily that's research and development. So things such as R and D tax credits are very important. If you look to industry, industries more and more are pulling back on research. Economists will argue how much government funding of research and development leads or follows corporate research and development. I'm inclined to think it leads. But either way, we need more of it.

Take energy research. We're spending more than $500 billion a year on energy and we're not spending 1 percent of that on research into alternatives. Whether you're a cobbler or a tinman if you see that your raw materials are in jeopardy, you extend some effort to find new sources. The point is that our supplies are in jeopardy, and our means of producing energy are environmentally damaging and yet we're paying little attention to alternatives.

— *We've heard you'll be advocating reforms in science education.*

Well, yes. Science education should not just be for future scientists, but for all kids. We should get away from this idea that in 10th grade you do biology, in 11th chemistry, and if you stick it out long enough, you do physics in the 12th. It should be all mixed up. And we should be teaching science in a way that's integrated with other disciplines—literature, social studies.

— *How did you get interested in science?*

I think when I was in elementary school a book on dinosaurs and fossils got me interested, and I decided then I would be a scientist. My mother was a biology teacher and I'd borrow her books. She had a master's in zoology. After my father died, she taught at a junior college for a couple of years. Then, later, she became Secretary of State in West Virginia, and she never got back to science. But while she was still teaching, I often went to the building where my mother was teaching and played around with the chemicals.

I didn't know at the time how dangerous it was to play with mercury. But it sure was fun.

— *Back to your recent campaign. When you heard Congressman Pappas singing "Twinkle, twinkle, Kenneth Starr" on C-Span, what was your first reaction?*

That we should rethink our media budget. I thought, every voter in the district should have the opportunity to hear that.

— *Would you say, if you were rewriting Machiavelli's* The Prince, *one of your new rules might be "Never sing on C-Span"?*

I think Al D'Amato also sang. What was it? "Old MacDonald Had a Farm." And I think Charles Schumer took a page from our campaign with resurrecting an old Al D'Amato song. And he's now Senator-elect Schumer.

November 24, 1998

Postscript

Any follower of political news knows that Rush Holt was returned to Congress by a hair-thin margin in November of 2000.

In one of those unfortunate early calls that television broadcasters are certain to be more cautious about in the future, CBS News even declared his opponent the victor. As with the presidential election, it was weeks before an official tally could be certified.

"Of the various things I've done in my life, my first term in Congress was the most challenging—intellectually, emotionally, and even, with all the travel and long hours, physically," Representative Holt wrote me on March 15, 2001.

"It was also, by far, the most satisfying. And it also seemed to be generally regarded as successful, by people both in New Jersey and in Washington. "I and my staff devoted our attention to the needs, problems, and concerns of my constituents in New Jersey. The issues I focussed most on were education (especially science education), health care, the environment, and research and development. And, throughout everything, I tried to give people reason to have more trust in their government.

"Based on my legislative success and my reputation for constituent service, I ran for re-election. It was clear from the beginning that I would face a strong challenge. A lot of political insiders hadn't expected me to win the first time, in 1998. Within hours of that victory, I had the rare distinction of having two former members of Congress announced to run against me. The one who ended up getting the Republican nomination had gotten 68 percent and 64 percent of the vote the last times he had been elected to Congress in my district. So I was in the cross hairs from the beginning. The conventional wisdom of the pundits in Washington was that I was the most vulnerable Representative in Congress. But over two years, I stayed in close touch with the people of New Jersey about what I was doing for them, and in the campaign the very real differences between me and my opponent came out. With help from environmentalists, teachers, labor organizations, advocates for women's health, and others, including some scientists, we could to identify my supporters and we worked hard to encourage them to get out to vote.

"The returns on election night gave me a narrow lead. Over the next few weeks, as my opponent challenged various kinds of ballots and precinct counts, my lead kept growing. He conceded when my margin grew to 738 votes, out of nearly 300,000. It was like what happened in Florida, but, on both sides, it was seen as more civil and dignified. "In the new Congress I've picked up where I left off at the end of last year. I have the same focus, I'm putting in the same effort, and it's still both challenging and rewarding."

ELEANOR BAUM

Bringing Feminine Mystique to Engineering

In a time when the words affirmative action are unmentionable in some corners of academia, **Eleanor Baum,** electrical engineer, dean of engineering at Cooper Union in New York, and the first woman to head an engineering college anywhere, is running a one-woman affirmative action campaign that is transforming engineering.

Since coming to Cooper Union 12 years ago, Dr. Baum, 59, has deliberately moved the female engineering enrollment at her tuition-free institution to 38 percent, from 5 percent.

Beyond her Cooper Union accomplishments, Dr. Baum has a history of breaking glass ceilings. She is chairwoman of the Board of Governors of the New York Academy of Sciences, and a director of Allegheny Power Systems, the United States Trust Company, and the Avnet Corporation.

Over coffee at her Cooper Union office on Astor Place on a recent morning, she talked about what led her to try to redesign the sexual composition of her profession.

___ *Why did you, in the midst of* The Feminine Mystique *years of the 1950s, become an engineer?*

Growing up in my family, you had to be such a "good" kid, so I took advanced math and science courses at Midwood High School in Brooklyn. And I was the only girl in the classes. The boys were all talking about becoming engineers. The girls at Midwood talked about marriage and teaching. I didn't like that.

I remember someone asking me, "Oh, what do you want to do?" and my answering "engineer." Their reaction was pure horror. "But people will think you're weird and no one will want to marry you," my mother gasped when I told her of my ambitions. At that point, I dug in my heels.

___ *Engineering school—City College—must have been a strange experience.*

229

Very. Because people really did think I was weird. I felt very conspicuous, very alone, but I finish what I start.

Some teachers were supportive; many were not. I was even hit on by
teachers. I remember taking a lab where the person teaching it was a married guy who asked me out. I was able to convince him it was beneath his
dignity to date a "mere" student.

I was so conspicuous. Every time I got back a test, everyone was terribly interested in my grade. And the hard part was the feeling that I had to
be "all women." If I didn't understand something, it meant that all women
didn't understand this.

— *What was your first job like?*

The first real job was at Sperry. It was awful, boring. People often
assumed that I was a secretary and not an engineer. When I had to consult
with another engineer, the thought was that I was over there flirting with
them. The supervisor had a talk with me: "I don't want you going over
there and flirting with the guys."

"I'm getting information for the project," I answered to his disbelief.

I lucked out, though. Brooklyn Polytech was looking for qualified
people they could offer graduate fellowships to. They called City College
and the departmental secretary there, a wonderful dear woman, gave them
my name. I had to teach one course. And I discovered that I loved teaching. I had become an engineer to avoid being a teacher, and here I was a
teacher, but I adored it.

— *The other evening here at the Cooper Union you had an Engineering
Career Night for high school girls. The main message seemed to be, "Young
woman, you're not a misfit if you want to be an engineer."*

Yes. Because the women I met when I was in college who were engineers were very strange ladies. Most of them were not married and had
given their lives to this profession and were very angry people.

I think that in these new times, you can have a career in engineering
and have a very normal life. Modern engineering, after all, is done in
teams. And people skills and team skills are terribly important. One reason I think it's important to have women go into this is that women do
have these skills. They also sometimes see problems very differently than
men. And this mix is crucial.

Interestingly, we did a survey of women engineer students and we found that more than two thirds of women who become engineers have a brother or a father who is one. And the reason is you have to know an engineer to understand what they do. Because what engineers do seems to be so mysterious.

— *What do engineers do?*

They solve problems, many of them technical in nature. What they really do is that simple—they make life better for people. They apply science toward this end. Those parts of engineering where that message is very clear are very attractive to women.

— *When you first came to Cooper Union more than a decade ago, did you tell the board that you intended to drastically change the gender makeup of the student body?*

No, I didn't. That's the wrong way to go about things. The right way is to get the job and then just do it. When I first came, there were 5 percent women. It's 38 now. And I love it. One of my triumphs is walking up the back stairs and finding necking couples, and they are both engineers.

— *How did you change the school's balance between men and women in engineering?*

I let it be known that I'd really like to look for women and minority faculty. I encouraged people to stop practices that make women uncomfortable. We've not rehired a couple of people who made remarks like, "You know, women don't really belong here."

I remember getting those remarks when I was a graduate student. I remember when I passed my doctoral-qualifying exams, the remarks of some of the men who didn't get passed were: "It's so unfair when she's only going to get married and have kids."

— *How has industry responded to all these female graduates you're suddenly turning out?*

Oh, industry loves it. In engineering, there are a lot of companies who get government contacts with affirmative action guidelines and they love the fact that there are women to hire.

— *When you were a very young child, in the middle of World War II, your parents took you and escaped Poland, traveled the width of Asia to get to Japan and then immigrated to Canada and eventually, the United States. Does some of your drive come from being a child survivor?*

Well, I don't think of myself as a child survivor. I think of my parents as the survivors. What drives me is being an only child and the intensity of my parents' expectations. It was only me. And with their having lost everybody in Europe, we were a very close unit. We were, in my mother's view, all that she had in the world. They wouldn't leave me with a baby sitter and took me everywhere.

The other piece of it, perhaps a less comfortable piece, is that both of my parents had accents. It was a time in American history when people with accents weren't trusted. They weren't "American," and it took a very, very long time for my parents to assimilate so I always grew up thinking my parents were "different." And a little strange. While I was proud of their achievements, there was an undercurrent of being a little uncomfortable. I wanted to be like everyone else. For example, I made a big point of forgetting Russian and Polish completely so that I could be a "real" American and only speak English.

— *On a less serious subject, your husband is a physicist. Do you agree with Rita Colwell, the head of the National Science Foundation, who believes that physicists make the best mates for female scientists?*

Well, it's true. I did get married. My mother was wrong. As for physicists, they are certainly not as pompous as many physicians are. They don't have a God complex. I think they realize that the universe is a complicated place.
June 22, 1999

Postscript

I met Eleanor Baum on an evening in the winter of 1999 when I went to the Cooper Union of New York to hear a world-famous mathematician give a talk about art.

This mathematician was someone I'd been considering for an interview, but his talk was over-the-top boring. And *pretentious*!!! When the gentlemen projected onto a screen his rather bad childhood doodles, I quietly laughed and scratched him off my list as a potential interviewee. What saved the evening was a post-speech dinner, where I met Dr. Baum, the Dean of the Cooper Union Engineering School; we stood in our high heels on the buffet line and she made the wait easier by telling marvelous stories about how she was feminizing that bastion of masculinity the profession of Engineering. *Tra-rah,* I had an interview after all.

Two years after the interview, in February of 2001, Dean Baum, sent this update on her most recent activities:

"I have been very involved in both national and international engineering matters. We've been trying to change engineering curricula so that graduates will have 'soft skills' in addition to their technical skills. We want engineers to be good communicators, have team skills, be aware of the global nature of the profession, understand the environmental and societal effects of their design decisions, and practice in an ethical manner.

"All of this seems obvious, but changing anything in universities is a very slow process. I've also been chairing a group called the Washington Accord. It is an agreement among most of the English-speaking nations dealing with easing mobility of engineers across borders.

"Finally, I've also been chairing the Engineering Workforce Commission, which deals with issues like the very real shortages projected for engineers and IT (information technologies) professionals. The only solution to that is to make engineering attractive as a career choice to women and minorities."

GEORGE E. BROWN JR.

The Congressman Who
Loved Science

Carol T. Powers

Representative **George E. Brown Jr.** of California, the ranking Democrat on the House Committee on Science and the former chairman of the House Committee on Science, Space and Technology, is dean of the Congressional science buffs.

Mr. Brown, 79, has been a crucial supporter of manned and unmanned space exploration. He was an author of legislation creating the Environmental Protection Agency, a prime mover behind efforts to include ozone layer protections in the Clean Air Act, and an advocate of restructuring the national weapons laboratories to meet the needs of a peacetime economy.

"From my earliest days, I was fascinated by science," Mr. Brown said on a recent afternoon in his offices on Capitol Hill, his wife, Marta, at his side. "Also by my earliest days, I was fascinated by a utopian vision of what the world could be like. I've thought that science could be the basis for a better world, and that's what I've been trying to do all these years."

— *How skilled are scientists and researchers at presenting their case to Congress?*

Very unskilled. They, generally speaking, have too great a faith in the power of common sense and reason. That's not what drives most political figures, who are concerned about emotions and the way a certain event will affect their constituency. If you're going to work in a political environment, you have to know the reasoning of the people you're dealing with. You have to talk to them realistically. It does very little good to appeal to high principle, although I would not say that's insignificant. The vast majority of politicians think they are functioning on high principle.

— *Some months ago, we interviewed Leon Lederman, the former head of the Fermi National Laboratory. It was his view that it may have been a mistake*

over the years for the scientists to ask for funding as a part of the cold war. Do
you agree with him?

Well, I wouldn't describe it as a complete mistake. But to build the funding of science for the next generation on the basis of the cold war was not
well advised. That implied that science wasn't important enough to survive
without a cold war. The truth is that science—research and development—
is probably the most important factor in the progress of the human race
over the last several thousand years. To base support of science on something that ephemeral as temporary alliances and enemies is ridiculous!

— *What's your take on the Czech president and playwright Vaclav Havel's*
famous statement that the fall of Communism has meant "the end" of science?

Well, like many people, I've tried to understand Mr. Havel as much as
I could. I think his feeling was that science was the ultimate expression of
a neo-modernist philosophy and that it is inherently bad because it fails
to consider human emotions and needs. It was probably his hope that
once the cold war was over and the need to have overwhelming military
strength had dissipated, that people would revert to a more benign and
caring and nonrational way of interacting.

Havel was influenced by the fact that the Communist system considered itself to be the ultimate in "scientific rationalism." Communism was
touted from the beginning as a "scientific" way of looking at the problems
of society. It wasn't. It probably never could be. It represented an effort to
control society with a philosophy that was antithetical to freedom and
democracy.

— *You're a strong advocate of international cooperation in science and yet,*
you are very critical of our State Department. Why?

Because the State Department gives lip service to the importance of
international cooperation in science. And lip service is about all they give.
They are reducing the number of scientific attachés at overseas missions,
and scientists in policy-making positions in the State Department here in
Washington.

My problem with the State Department is that they spend most of
their time in this area in long, drawn-out international negotiations. They
negotiate a new science and technology cooperation treaty. You can

almost count on their doing that periodically with each country we have good relations with. Basically, it's a photo opportunity for the presidents of the two countries. Then, believe it or not, the administration doesn't request any money to carry out the provisions of the agreement. This isn't unique to the current administration; they all do it.

— *The microbiologist Rita Colwell, an expert on cholera, was named last year to be the nation's top scientist, to head the National Science Foundation. Does her appointment represent a sea change in the NSF's direction?*

Her appointment could portend a new direction in the support of science in the United States. Her career and her achievements reflect the direction science funding is going into in the future. In other words, jobs, industrial opportunities are going to stem more from the biological sciences than from chemistry and physics. I see biology as being the greatest area of scientific breakthroughs in the next generation. So, I'll probably be supporting her in any changes she wants to make in the direction of the N.S.F.'s funding.

— *Are you a scientist?*

I'm not. I'm a politician. I started out with a background in science. I do have a degree in physics and did a short stint as an engineer. I was interested in science before I even knew what science was. I used to read science-fiction magazines when I was a kid, stories about space travel, and that led me—structuring my college training in math, physics, chemistry, astronomy—a little bit. I've been on the Science Committee in the House since 1965.

— *When you are putting together science policy, how much do you rely on staff?*

How do you quantify that? I rely very, very heavily on the staff of the House Science Committee. That's because we have some of the best people with the best scientific backgrounds available anywhere in the country. These people are good enough to know how to express my views, even when they don't agree with them. That's always very helpful.

— *Former Senator John Glenn was able to parlay his support for science into a second ride in space. Have you considered doing the same?*

(Laughs) I've indicated to NASA Administrator Dan Goldin that if he thinks I'm physically qualified, I'd still like to go into space. In any event, politicians have been known to use their influence to get things that they want.

—— *When you give speeches, you say that you expect to watch the progress of the space program from your perch in heaven some day. What exactly do you mean?*

That I think we're soon going to have the exploration of the heavens by humans, starting with Mars and the rest of the solar system. I think maybe Dan Goldin and a few other people have this sense that human settlement of the universe is, beginning with what's nearest to us, an inevitable progression of human beings.

—— *Where do you vote on the "is there other life out there" question?*

I consider the probability that there is other intelligent life in the universe overwhelming. And we will ultimately find it. It may not be exactly what we think it ought to be, but it will be intelligent life. It may be far more intelligent than we are. And that will be an interesting situation. How do we deal with a situation where we are encountering someone with vastly superior intellectual powers to ourselves? Now, I'm not fearful of that. Because highly evolved life will not be looking at new life forms as enemies. They will be looking on them as their children who need to be encouraged to develop even further.

—— *Do you see a beneficent kind of* Star Trek *scenario for our space future?*

I think it will be beneficent, and I like *Star Trek*. The people who write *Star Trek* are going through all these same scenarios as you are. They are asking, "What is a reasonable rational future for the human race?" And then they try to project that into a television program.

—— *I don't mean this disrespectfully, but as a senior member of the House, do you ever think about retirement?*

I have no desire or ambition to enter another career just to earn more money or something like that. So my criteria for retirement will be determined by my sense of whether or not I'm adequately fulfilling the responsibilities I have.

Some of my colleagues, years before they retired, weren't accomplishing anything. I don't want to do that. What I've always wanted to do is help shape ideas about the emerging human culture, the kind of thing that Newt Gingrich sometimes did. I think I can make a small contribution to that, maybe not as much as Newt. But I think I'm more on the right track. We'll see. As to my decision as to whether I decide to retire or not, that may be conditioned on how you write this story.
March 3, 1999

Postscript

Congressman George Brown did not seem particularly well when I interviewed him in January of 1999, at the American Association for the Advancement of Science annual convention in Anaheim, California.

He had a lingering cold, he complained. The endless airplane trips between Washington and home district in southern California were aggravating his condition.

And so we had lunch and did an interview in my room at the Anaheim Hilton—myself, the Congressman, and his wife, Marta Marcias Brown, who was also his Executive Assistant and traveled everywhere with him.

Ours was a genial lunch. But Mr. Brown was bone-tired and the interview proved bland and uninteresting. As I listened to the tapes later, I felt that none of the vibrancy of this eighteen-term Congressman had come through. Thus, as soon as I returned to the east coast, I asked for another session. One of the joys of doing these sorts of conversations is that one can sometimes go back for a second take.

Happily, the second interview proved more exciting. We taped in his wood-paneled offices in the Rayburn House Office Building in Washington. The ever-loyal Marta was there, as were several of Brown's legislative aides.

Less than five months later the Congressman was dead.

"George had congestion in his lungs when you saw him," one of his former aides, Dan Pearson, told me on the telephone in March of 2001. "He wasn't getting enough oxygen, and they were treating the symptoms. He finally went in March for heart surgery. The operation went well. He returned to work and was O.K. Nine or ten days later, he felt awful. He had an infection from the surgery. . . . They put a plastic (heart valve) one in and began treating the infection. He got through that. A staph infection set in. He didn't survive it. He was 79."

In March of 2001, I spoke with Marta Brown in her office in California. I wanted to confirm what I'd learned elsewhere. Quite naturally, Mrs. Brown spoke with pained difficulty of the events that had made her a widow. "It was one of those things that happens," she sighed. Mrs. Brown is currently heading a foundation to build a science education center in George Brown's memory in his home district, in Loma Linda, California. "He had a heart valve replacement and ran into difficulties with that. In spite of his work in the field of science, science couldn't save him."

And here, with George Brown's death, is one of the ironies surrounding modern science. Through we live in a time when sheep can be cloned and the genome has been mapped, the simpler lessons that previous generations of scientists have taught us are often ignored to our peril. Staph infections are often contracted in hospitals after surgery. It's hard to forget the voice of one of Representative Brown's Congressional friends, who believed that "George died because someone didn't wash their hands."

An Idiosyncratic Bibliography of Science Reading for the Novice

Angier, Natalie. *Woman: An Intimate Geography.* New York: Houghton Mifflin Company, 1999. The Pulitzer Prize–winning journalist's investigation of "What makes a woman?"

Altman, K. Lawrence, M.D. *Who Goes First? The Story of Self-Experimentation in Medicine.* Berkeley and Los Angeles: University of California Press, 1998.

Atkins, P. W. *Periodic Kingdom: A Journey into the Land of Chemical Elements* (Science Master Series). New York: HarperCollins, 1997. It's elementary.

Balick, Michael J., and Paul Alan Cox. *Plants, People, and Culture: The Science of Ethnobotany.* New York: Scientific American Library, 1996. Balick and partner's gorgeously illustrated introduction to their line of work.

Bolles, Edmund Blair, Ed. *Galileo's Commandment: 2500 Years of Great Science Writing.* New York: W. H. Freeman and Company, 1999. 2500 Years of Inspiration.

Borsook, Paulina. *Cyberselfish.* A former contributing writer to *Wired Magazine* gives an iconoclastic report on the downside of the e-culture.

Brody, E. Jane, and the reporters of *The New York Times,* Denise Grady, Ed. *The New York Times Book of Women's Health.* New York: Lebhar-Friedman Books, 2000. An important collection of reporting on the everyday health issues that women confront.

Burrows, William E. *This New Ocean: The Story of the First Space Age.* New York: The Modern Library, 1999. The science educator's epic true life tale of space exploration during the era of the cold war.

Burrows, William E. *Deep Black: Space Espionage and National Security.* New York: Random House, 1987.

Carson, Ben S., with Cecil Murphey. *Gifted Hands, The Ben Carson Story.* Grand Rapids: Zondervan Publishing House, 1990. Dr. Benjamin Carson's spiritual memoir.

Casti, John L. *Paradigms Regained: A Further Exploration of the Mysteries of Modern Science.* New York: William Morrow, 2000.

Cavalli-Sforza, Luigi Luca. *Genes, Peoples, and Languages.* New York: North Point Press, 2000. The great geneticist sums up a lifetime of investigation into whether or not human genes contain a historic record of homo sapiens.

Chang, Laura, Ed. *Scientists at Work: Profiles of Today's Groundbreaking Scientists from Science Times.* New York: McGraw-Hill, 2000. An excellent collection of profiles from the popular *New York Times* series. Particularly useful for journalism school students studying science writing and for young people considering a scientific career.

Coen, Enrico. *The Art of Genes: How Organisms Make Themselves.* New York: Oxford University Press, 1999. Dr. Coen links art and genetics.

Danielson, Dennis Richard, ed. *The Book of the Cosmos: Imagining the Universe from Heraclitus to Hawking.* Cambridge: Perseus, 2000.

Darwin, Charles. *The Voyage of The Beagle.* Edited by Leonard Engel. New York: National History Library, 1962. In the beginning, there was Darwin and his long brave journey to the other side of the earth and to human knowledge of our origins. This is his adventure story in his own words.

Dawkins, Richard. *The Extended Phenotype: The Long Reach of the Gene.* Revised Edition. New York: Oxford University Press, 1976.

Dawkins, Richard. *The Selfish Gene.* New York: Oxford University Press, 1989. The king of the neo-Darwinists explains his basic ideas.

Dean, Cornelia. *Against the Tide: The Battle for America's Beaches.* New York: Columbia University Press, 1999. The Editor of the Tuesday *Science Times* section of *The New York Times* shows how unwise human intervention is destroying our nation's beachfronts. Her point: "In the long run, extensive development cannot coexist with an eroding beach—and most American beaches are eroding."

Dertouzos, Michael L. *What Will Be: How the New World of Information Will Change our Lives.* New York: HarperCollins, 2001.

Dertouzos, Michael L. *The Unfinished Revolution: Human-Centered Computers and What They Can Do for Us.* New York: HarperCollins, 2001.

Devlin, Keith. *The Language of Mathematics: Making the Invisible Visible.* New York: W. H. Freeman and Company, 2000. Indispensable for math klutzes.

Dubos, Rene. *The God Within.* New York: Scribner, 1984.

Dubos, Rene. *Louis Pasteur, Free Lance in Science.* Cambridge: Da Capo, 1986.

Dyson, Freeman. *The Sun, the Genome and the Internet.* New York: Oxford University Press, 1999. Thought-provoking essays on the important scientific issues of our time.

Dyson, Freeman. *Origins of Life.* Second Edition. New York: Cambridge University Press, 1999.

Dyson, Freeman. *Disturbing the Universe.* New York: Harper & Row, 1979. Dr. Dyson's very undisturbing memoir.

Eisely, Loren C. *The Immense Journey.* New York: Random House, 1959.

Eldredge, Niles. *Life in the Balance: Humanity and the Biodiversity Crisis.* Princeton: Princeton University Press, 1998.

Faigman, David L. *Legal Alchemy: The Use and Misuse of Science in the Law.* New York: W. H. Freeman and Company, 2000.

Fausto-Sterling, Anne. *Sexing the Body.* New York: Basic Books, 2000. A further development of the work begun fifteen years earlier with her previous book, *Myths of Gender.*

Fausto-Sterling, Anne. *Myths of Gender: Biological Theories About Women and Men.* New York: Basic Books, 1985. This is the book that provided some scientific answers to those who swore that "biology is destiny."

Feynman, Richard P. *The Meaning of It All: Thoughts of a Citizen-Scientist.* Reading, PA: Perseus, 1998. The wittiest and most brilliant man in American physics uttered these words in a 1963 lecture series. They still ring true today.

Feynman, Richard P. *The Pleasures of Finding Things Out: The Best Short Works of Richard P. Feynman.* Edited by Jeffrey Robbins. Cambridge: Perseus, 1999. Essays and lectures by the 1965 Nobel Prize winner in physics that show why the late Feynman was one of the great scientists of the twentieth century. There is a moving foreword by his old friend, Freeman Dyson.

Fiffer, Steve. *Tyrannosaurus Sue: The Extraordinary Saga of the Largest, Most Fought Over T. Rex Ever Found.* New York: W. H. Freeman and Company, 2001. How "Sue" got installed at an elegant address right on Lake Michigan in Chicago.

Fischer, Ernst Peter. *Beauty and the Beast: The Aesthetic Moment in Science.* Translated by Elizabeth Oehlkers. New York: Plenum Trade, 1999. A German historian of science examines scientific discovery and beauty— and does it in a most interesting way.

Flatow, Ira. *They All Laughed . . . From Light Bulbs to Lasers: The Fascinating Stories Behind the Great Inventions That Have Changed Our Lives.* New York: Harper Perennial, 1993. Broadcaster Flatow shows how science is all around us and how the greats are those who see what was always there.

Galdikas, Birute, and Nancy Briggs. *Orangutan Odyssey.* Photographs by Karl Ammann. New York: Harry N. Abrams, Inc., 1999. Dr. Galdikas's gorgeous photo memoir of her life with the great apes of Borneo.

Galdikas, Birute. *Reflections of Eden.* New York: Little, Brown and Company, 1995.

Galilei, Galileo. *Two New Sciences.* Translated by Stillman Drake. Second Edition. Dover: Adasi, 1989.

Gell-Mann, Murray. *The Quark and the Jaguar: Adventures in the Simple and the Complex.* New York: W. H. Freeman and Company, 1995. This is an introduction to the complex mind of Dr. Gell-Mann, one of the great thinkers of our time.

Gillies, James, and Robert Cailliau. *How the Web Was Born.* New York: Oxford University Press, 2000.

Goldsmith, Donald. *The Runaway Universe: The Race to Find the Future of the Cosmos.* Cambridge: Perseus, 1999.

Goodall, Jane, with Phillip Berman. *Reason for Hope: A Spiritual Journey.* New York: Warner Books, 2000. Jane Goodall seeks to make the case for science and faith.

Gosden, Roger. *Designing Babies: The Brave New World of Reproductive Technology.* New York: W. H. Freeman and Company, 2000.

Gratzer, Walter. *The Undergrowth of Science: Delusion, Self-Deception, and Human Frailty.* New York: Oxford University Press, 2000.

Greene, Brian. *The Elegant Universe: Superstrings, Hidden Dimensions, and the Quest for the Ultimate Theory.* New York: Vintage, 2000.

Hall, Stephen S. *A Commotion in the Blood: Life, Death, and the Immune System.* New York: Henry Holt and Company, 1997. Stephen Hall's impressive investigation of the revolutionary new immunotherapies that are changing everything about the way we treat AIDS, cancer, and many viral diseases.

Hawking, Stephen. *A Brief History of Time.* New York: Bantam Doubleday Dell Publishing, 1998. Hawking's monumental work, not brief, not easy, but worth it.

Hawkins, Michael, with Celia Fitzgerald. *Hunting Down the Universe: The Missing Mass, Primordial Black Holes, and Other Dark Matters.* Reading, PA: Perseus, 1997.

Hellman, Hal. *Great Feuds in Medicine: Ten of the Liveliest Disputes Ever.* New York: John Wiley & Sons, Inc., 2001. For anyone who thought the political world alone is brutal, this probe into the visciousness of medical feuds is extremely edifying.

Horgan, John. *The End of Science: Facing the Limits of Knowledge in the Twilight of the Scientific Age.* New York: Broadway Books, 1997. Horgan's influential snarl.

Horgan, John. *The Undiscovered Mind: How the Human Brain Defies Replication, Medication, and Explanation.* New York: The Free Press, 1999. Progress is not inevitable, says John Horgan.

Kaplan, Robert. *The Nothing That Is: A Natural History of Zero.* New York: Oxford University Press, 2000. The something that is: Robert Kaplan's wonderful exploration.

Kemp, Martin. *Visualizations: The NATURE Book of Art and Science.* Berkeley and Los Angeles: University of California Press, 2000. This beautifully illustrated volume shows the reader all the places where art and nature inspire, combine, and cross-fertilize. The author is a *NATURE* columnist.

Kolata, Gina. *Flu: The Story of the Great Influenza Pandemic of 1918 and the Search for the Virus That Caused It.* New York: Farrar, Straus and Giroux, 1999.

Kolata, Gina. *Clone: The Road to Dolly and the Path Ahead.* New York: William Morrow, 1998

Larson, Edward J. *Evolution's Workshop: God and Science on the Galapagos Islands.* New York: Basic Books, 2001. This should be perused after a rereading of *Voyage of the Beagle.*

Leakey, Richard E. *One Life: An Autobiography.* Salem: Salem House, 1984. Leakey's life story, which continues to have fascinating chapters, is riveting. His is one of the major scientific adventure stories of the twentieth century.

Lederman, Leon M., and David Schramm. *From Quarks to the Cosmos: Tools of Discovery.* New York: Scientific American Library Paperbacks, 1989.

Lederman, Leon, with Dick Teresi. *The God Particle: The Universe Is the Answer, What Is the Question?* New York: Delta, 1993. The "Mel Brooks of physics," writes a funny book about creation, particle physics, and some of the fun he's had as a citizen-physicist running Fermilab.

Levin, Simon. *Fragile Dominion: Complexity and the Commons.* Reading, PA: Perseus, 1999.

Lewis, John S. *Worlds Without End: The Exploration of Planets Known and Unknown.* Reading, PA: Perseus, 1998.

Maeda, John. *Maeda @ Media.* New York: Rizzoli, 2000. Georgeous compilation of Maeda's recent work.

Maeda, John. *Design by Numbers.* Cambridge: The M.I.T. Press, 1999. Maeda's own computer language designed for artists. Or vice-versa?

Miller, Arthur I. *Einstein, Picasso: Space, Time, and the Beauty That Causes Havoc.* New York: Basic Books, 2001.

McEvoy, J.P., and Oscar Zarate. *Introducing Stephen Hawking.* Edited by Richard Appignanesi. New York: Totem Books, 1995. Part of a brilliant and beautiful comic book series that can only be described as "Cliff Notes meets *Mad* Magazine."

McEvoy, J.P., and Oscar Zarate. *Introducing Quantum Theory.* Edited by Richard Appignanesi. Cambridge: Icon Books UK, 1999. This witty comic book was indispensible for a certain science writer when she first started on her beat. It's probably plenty useful to high school and college students in a bad physics jam, too.

McGrayne, Sharon Bertsch. *Nobel Prize Women in Science: Their Lives, Struggles and Momentous Discoveries.* Revised Edition. Secaucus: Citadel Press, 1998. There are big life lessons in all their stories.

McPhee, John. *Annals of the Former World.* New York: Farrar, Straus and Giroux, 1998. This work proved that geologists do lead interesting lives. It also won a Pulitzer Prize.

Minsky, Marvin L. and Harry Harrison. *The Turing Option.* New York: Warner Books, 1992. Marvin Minsky tries his hand at a futuristic novel.

Minsky, Marvin L. *The Society of Mind.* New York: Simon & Schuster, 1988. This is Dr. Minsky's classic work. Hard to find. But worth it.

Minsky, Marvin L., with Seymour A. Papert. *Perceptrons.* Cambridge: M.I.T. Press, 1969.

Newbold, Heather, Ed. *Life Stories: World-Renowned Scientists Reflect on Their Lives and the Future of Life on Earth.* Berkeley: University of California Press, 2000.

Olshansky, S. Jay, and Bruce A. Carnes. *The Quest for Immortality: Science at the Frontiers of Aging.* New York: W. W. Norton & Company, 2001. A topic of great interest to every reader.

Orbach, Susie. *The Impossibility of Sex: Stories of the Intimate Relationship Between Therapist and Patient.* New York: Scribner, 2000. Fictionalized case studies from Dr. Orbach's London practice.

Overbye, Dennis. *Einstein in Love: A Scientific Romance.* New York: Viking, 2000. The personal Einstein during his most productive years is revealed as a horrid husband and father, but also as a man who changed everything we know about the universe. Ten years worth of research have netted a biography that is at once admiring and critical—Albert Einstein as cad and genius.

Overbye, Dennis. *Lonely Hearts of the Cosmos: The Story of the Scientific Quest for the Secret of the Universe.* New York: Back Bay Books, 1999. Dennis Overbye's vivid telling of the human dramas behind the big contemporary physics discoveries.

Peat, F. David. *The Blackwinged Night: Creativity in Nature and Mind.* Reading, PA: Perseus, 2000.

Peebles, Curtis. *Asteroids: A History.* Washington: Smithsonian Institution Press, 2000. Who could resist a work about asteroids penned by a writer named Peebles?

Peterson, Dale, and Jane Goodall. *Visions of Caliban: On Chimpanzees and People.* New York: Houghton Mifflin Company, 1993. Essays from Jane Goodall on her continuing effort to study and save the chimps of Gombe Stream and the world.

Quammen, David, Ed. *The Best American Science and Nature Writing 2000.* Series Editor, Burkhard Bilger. New York: Houghton Mifflin Company, 2000. One of our best nature writers picks the winners among his peers.

Raymo, Chet. *365 Starry Nights: An Introduction to Astronomy for Every Night of the Year.* New York: Simon and Schuster, 1992. Trade your astrological chart for this one.

Rees, Martin. *Before the Beginning.* Cambridge: Perseus, 1997. One of the great living astronomers explains the universe in simple and clear language. This may be "cosmology for dummies," but it is not dumb.

Rees, Martin. *Just Six Numbers.* Cambridge: Perseus, 2000. Sir Martin explains to us new discoveries about the universe and why those little six numbers determine "the essential features of the physical cosmos."

Rees, Martin, with Mitchell Begelman. *Gravity's Fatal Attraction.* New York: Scientific American Library 1996.

Ridley, Matt. *Genome.* New York: HarperCollins, 2000. Useful guide to the revolution of our times.

Sagan, Carl. *Cosmos.* New York: Random House, 1983. Though almost "billions and billions" of copies of this picture book have sold over the years, it remains one of the best introductions to astronomy, Ever.

Sapolsky, Robert M. *A Primate's Memoir.* New York: Scribner, 2001. The primate in question is Dr. Sapolosky himself. He is also an influential baboon researcher.

Sapolsky, Robert M. *Why Zebras Don't Get Ulcers: An Updated Guide to Stress, Stress-Related Diseases, and Coping.* New York: W. H. Freeman and Company, 1998. If zoology married self-help, this would be their offspring.

Sardar, Ziauddin, and Iwona Abrams. *Introducing Chaos.* Edited by Richard Appignanesi. Cambridge: Icon Books UK, 1999. Not chaotic, though.

Savage-Rumbaugh, Emily Sue, Stuart G. Shanker, and Talbot J. Taylor. *Apes, Language, and the Mind.* New York: Oxford University Press, 1998. An academic work on Savage-Rumbaugh and colleagues' recent experiences working with the Georgia State Universities bonobos.

Savage-Rumbaugh, Emily Sue, and Roger Lewin. *Kanzi: The Ape at the Brink of the Human Mind.* New York: John Wiley & Sons, Inc., 1994. This marvelous book tells how Emily Sue Savage-Rumbaugh came to communicate with Kanzi and his relatives.

Schrodinger, Erwin. *Nature and the Greeks: And Science and Humanism.* New York: Cambridge University Press, 1996.

Schrodinger, Erwin. *What Is Life? The Physical Aspect of the Living Cell with Mind and Matter and Autobiographical Sketches.* New York: Cambridge University Press, 1992.

Sheldon, Jennie Wood, Michael J. Balick, and Sarah A. Laird. *Medicinal Plants: Can Utilization and Conservation Coexist?* New York: The New York Botanical Garden, 1997. Michael Balick and collegaues make a strong case that they can.

Shilts, Randy. *And the Band Played On: Politics, People and the AIDS Epidemic.* New York: St. Martin's Press, 1987.

Singer, Peter. *Animal Liberation: A New Ethics for Our Treatment of Animals.* New York: Avon Books, 1975. Jane Goodall once told me that this is one of the most important books in her library.

Smolin, Lee. *The Life of the Cosmos.* New York: Oxford University Press, 1998.

Sobel, Dava. *Longitude: The True Story of a Lone Genius Who Solved the Greatest Scientific Problem of His Time.* New York: Walker and Company, 1995. Former *Science Times* reporter Sobel tells the tale of John Harrison's forty-year obsession to invent a clock that could keep time at sea. This slim volume was a best seller for a good reason—it's a gripping read.

Tanzi, Rudolph E., and Ann B. Parson. *Decoding Darkness: The Search for the Genetic Causes of Altzheimer's Disease.* Cambridge: Perseus, 2000. Rudi Tanzi tells the insider's story of how scientists are fighting (and beginning to win) the battle against Alzheimer's.

Thomas, Lewis. *Lives of a Cell: Notes of a Biology Watcher.* New York: Penguin USA, 1974. This is the classic. Science writing was never better.

Tiger, Lionel. *The Decline of Males.* New York: Golden Books, 1999. The neo-Darwinist argument on gender presented by one of its foremost exponents.

Tudge, Colin. *The Variety of Life: A Survey and a Celebration of All the Creatures That Have Ever Lived.* New York: Oxford University Press, 2000.

Waal, Frans de. *The Ape and the Sushi Master: Cultural Reflections of a Primatologist.* New York: Basic Books, 2001. De Waal puts his prestige on the line to weigh in on the animal consciousness controversy.

Waal, Frans de, and Frans Lanting. *Bonobo: The Forgotten Ape.* Berkeley and Los Angeles: University of California Press, 1997. This is the absolute last grunt on those wonderful apes that Emily Sue Savage-Rumbaugh studies and that de Waal has come to admire.

Wade, Nicolas. *Lifescript.* New York: Simon and Schuster, 2001

Walker, Evan Harris. *The Physics of Consciousness: The Quantum Mind and the Meaning of Life.* Cambridge: Perseus, 2000.

Watson, James. *The Double Helix: A Personal Account of the Discovery of the Structure of DNA.* New York: W. W. Norton & Co., 1980. This classic never dates. It should be reread now that the genome has been mapped.

Weinberg, Steven. *Dreams of a Final Theory.* New York: Vintage, 1994.

Wells, Martin. *Civilization and the Limpet.* Reading, PA: Perseus, 1998, Fun reading by a witty zoologist. Squishy tales about squids, jellyfish, and other invertebrates.

Wilford, John Noble. *The Mapmakers: The Story of the Great Pioneers in Cartography—From Antiquity to First Revised Edition.* New York: Alfred A. Knopf, 2000. The reissue of the great science writer's truly great work on the history of cartography.

Wilford, John Noble. *Mars Beckons: The Mysteries, the Challenges, the Expectations of Our Next Great Adventure in Space.* New York: Vintage Books, 1991

Wills, Christopher. *Children of Prometheus: The Accelerating Pace of Human Evolution.* Reading, PA: Perseus, 1998. As his title suggests, Dr. Wills believes that the pace of human evolution is proceeding at an ever-increasing rate. Wills is one of those rare academics who writes well.

Wills, Christopher, and Jeffrey Bada. *The Spark of Life: Darwin and the Primeval Soup.* Cambridge: Perseus, 2000.

Wilmut, Ian, Keith Campbell, and Colin Tudge. *The Second Creation: Dolly and the Age of Biological Control.* New York: Farrar, Straus and Giroux, 2000.

Wright, Michael, and Mukul Patel, Editors. *Scientific American—How Things Work Today.*

Young, Louise B. *Islands: Portraits of Miniature Worlds.* New York: W. H. Freeman and Company, 2000. Scientist, traveler, and poet Louise B. Young visits the world's most fantastic islands and tells us exactly why they are so wonderful.

Zubrin, Robert. *Entering Space: Creating a Spacefaring Civilization.* New York: Tarcher Putnam, 1999.

Zubrin, Robert, with Richard Wagner. *The Case for Mars.* New York: Simon and Schuster. 1997. Robert Zubrin's influential exhortation for human exploration of Mars.

Zubrin, Robert. *First Landing.* New York: Ace Putnam, July 2001. Zubrin's first landing in the fiction zone.

Zuger, Abigail. *Strong Shadows: Scenes from an Inner City AIDS Clinic.* New York: W. H. Freeman and Company, 1997. Emotionally moving piece on a story whose drama, sadly, does not diminish with time.

Zukowsky, John, Ed. *2001: Building for Space Travel.* New York: Abrams, 2001. This companion volume to the Art Institute of Chicago's exhibition examines space travel design as seen from the artist's and the scientist's point of view. Lots of fascinating pictures of how popular culture has boldly dreamed of outer space.